Distinguished Dissertations

Springer

London
Berlin
Heidelberg
New York
Barcelona
Hong Kong
Milan
Paris
Santa Clara
Singapore
Tokyo

Masahito Hasegawa

Models of Sharing Graphs

Graphs

A Categorical Semantics of let and letrec

Springer

Masahito Hasegawa, PhD
Research Institute for Mathematical Sciences, Kyoto University
Kyoto 606-8502, Japan

Series Editor

Professor C.J. van Rijsbergen
Department of Computing Science, University of Glasgow, G12 8RZ, UK

ISBN-13: 978-1-4471-1221-1

British Library Cataloguing in Publication Data
Hasegawa, Masahito
    Models of sharing graphs : a categorical semantics of let
    and letrec. – (Distinguished dissertations)
    1.Graph grammars 2.Computable functions
    I.Title
    511.5
    ISBN-13: 978-1-4471-1221-1

Library of Congress Cataloging-in-Publication Data
Hasegawa, Masahito, 1970-
    Models of sharing graphs : a categorical semantics of let and
    letrec / Masahito Hasegawa.
        p.   cm. -- (Distinguished dissertations)
    Originally published as the author's thesis, Ph.D, University of
    Edinburgh.
        ISBN-13: 978-1-4471-1221-1   e-ISBN-13: 978-1-4471-0865-8
        DOI: 10.1007/978-1-4471-0865-8

    1.Programming languages (Electronic computers) 2. Type theory.
    3.Categories (mathematics) I. Title. II. Series: Distinguished
    dissertations (Springer-Verlag)
    QA76.7.H38 1999   99-25290
    005.13--dc21   CIP

Typesetting: Camera-ready by author

34/3830-543210 Printed on acid-free paper

To my parents

Preface

A general abstract theory for computation involving shared resources is presented. We develop the models of *sharing graphs*, also known as term graphs, in terms of both syntax and semantics.

According to the complexity of the permitted form of sharing, we consider four situations of sharing graphs. The simplest is first-order acyclic sharing graphs represented by let-syntax, and others are extensions with higher-order constructs (lambda calculi) and/or cyclic sharing (recursive letrec binding). For each of four settings, we provide the equational theory for representing the sharing graphs, and identify the class of categorical models which are shown to be sound and complete for the theory. The emphasis is put on the algebraic nature of sharing graphs, which leads us to the semantic account of them.

We describe the models in terms of the notions of symmetric monoidal categories and functors, additionally with symmetric monoidal adjunctions and traced monoidal categories for interpreting higher-order and cyclic features. The models studied here are closely related to structures known as notions of computation, as well as models for intuitionistic linear type theory. As an interesting implication of the latter observation, for the acyclic settings, we show that our calculi conservatively embed into linear type theory. The models for higher-order cyclic sharing are of particular interest as they support a generalized form of recursive computation, and we look at this case in detail, together with the connection with cyclic lambda calculi.

We demonstrate that our framework can accommodate Milner's *action calculi*, a proposed framework for general interactive computation, by showing that our calculi, enriched with suitable constructs for interpreting parameterized constants called controls, are equivalent to the closed fragments of action calculi and their higher-order/reflexive extensions. The dynamics, the computational counterpart of action calculi, is then understood as rewriting systems on our calculi, and interpreted as local preorders on our models.

Preface to the Present Edition

This book contains the author's PhD thesis written under the supervision of Rod Burstall (first supervisor), Philippa Gardner and John Power (second supervisors) at Laboratory for Foundations of Computer Science, University of Edinburgh. The thesis was examined by Martin Hyland (Cambridge) and Alex Simpson (Edinburgh). Except for correcting minor mistakes and updating the bibliographic information, the text agrees with the examined version of the thesis.

Some parts of the book have been published elsewhere in [13, 35, 38]. Since the examination of the thesis, a number of works related to this research have appeared. I take this opportunity to mention some of them.

- An independent work by Corradini and Gadducci [25] used essentially the same categorical structure described in Chapter 3 for modeling acyclic graph rewriting systems (with Cat-enrichment rather than Preord-enrichment). Miyoshi [70] translated the results in Chapter 6 to their setting and reformulated the cyclic sharing theories as a rewriting logic.

- While the model construction techniques in Chapter 5 show the conservativity of syntactic translations, further techniques for showing the fullness (or full completeness) of the translations have been developed by the author, as reported in [39].

- A direction progressing rapidly is the investigation of traced monoidal categories as a foundation of recursive computation, as claimed in Chapter 7. Some fundamental issues on traced monoidal categories are studied in Abramsky, Blute and Panangaden [4] and Blute, Cockett and Seely [23]; the latter contains a fixpoint theorem related to those in Chapter 7. As an interesting case study, Ryu Hasegawa [40] related the fixpoint operator in a model of (typed and untyped) lambda calculus and the Lagrange-Goodman inversion formula in enumerative combinatorics in terms of trace. The relation to axiomatic domain theory has been studied by Plotkin and Simpson [74].

- In Chapter 9 the possibility of developing the premonoidal variant of the sharing theories and their models was suggested. Related to this, Jeffrey [46] has introduced a semantics of the graphically-presented imperative programs based on premonoidal categories. In that setting, he also modeled recursion using trace.

Acknowledgements

I want to express my heartfelt thanks to my supervisors; I can never thank them enough.

Rod Burstall, my first supervisor, always helped me to think constructively and positively, especially at difficult moments throughout my PhD study in Edinburgh. I will never forget a meeting with Rod when I had a bad hangover – there I got an essential inspiration in deciding my research direction.

Philippa Gardner, my second supervisor during the second year, always gave me enthusiastic encouragement, and I benefited from countless delightful (often over-heated) discussions with her.

John Power has always been an important intellectual source and often a mentor for me during these three years, and he became my second supervisor after Philippa moved to Cambridge. Without his generous and much needed support, this thesis would probably never have been written in this form.

At a stimulating place like LFCS, even a brief chat often meant a lot to me. I am grateful to people who influenced me in various forms, especially to Andrew Barber, Ewen Denney, Marcelo Fiore, Alex Mifsud, Robin Milner, Gordon Plotkin and Alex

Simpson. In particular, Chapter 5 and Chapter 8 refer to joint work with Andrew, Philippa and Gordon.

I want to thank Martin Hyland for helpful discussions on traced monoidal categories as well as for his warm encouragement. Thanks are also due to Zena Ariola and Stefan Blom, for e-mail communications on cyclic lambda calculi and related topics.

And, above all, many, many thanks go to my cheerful, lovely, friends.

This work was partly supported by an Oversea Research Student award.

Contents

1
Introduction

1.1 Computation Involving Shared Resources

The notion of *sharing* has appeared on various occasions in computer science, either explicitly or implicitly. The idea is simple: instead of giving computational resources (processes, memories etc) to each client, a single resource can be shared by multiple clients.

In general, this kind of replacement may change the nature of the involved computation significantly. For instance, if the resource we are concerned with requires heavy computation or a large memory, sharing becomes an essential technique for saving both time and space needed for the computation. Many implementations of pure functional programming languages are based on this observation – avoiding unnecessary duplication of subcomputation is crucial for achieving efficient functional computation.

However, sharing is not just about the efficiency. If the resource involves some computation with side effects, say non-determinism or imperative states, the sharing of such a resource may change not just the amount of computation but also the result of computation. In such impure cases, the distinction between duplicated resources and shared resources must be made more carefully, and this makes it difficult, or at least non-trivial, to reason about general computation involving shared resources.

Furthermore, sharing can naturally be used for implementing *cyclic* (self-referential) data structures, which have been used for implementing recursive computation efficiently. The expressive power obtained by cyclic sharing is enormous, but dealing with cyclic structures is far more difficult than dealing with just acyclic ones. For instance, there are various practical ways of encoding recursive computation using cyclic sharing, but, to the best of our knowledge, there has been no formal comparison between them.

This thesis is devoted to giving a theory for describing and reasoning about such computation with sharing. The weight is put on the study of the *classes* of models of sharing, rather than individual specific models, in a desire to extract a generic account for sharing.

1.2 Sharing Graphs as Models of Sharing

Sharing for Efficiency

No programmer would be happy to write an expression like

```
... (factorial(100) + 123) * factorial(100) ...
```

containing two identical subexpressions `factorial(100)` – not just because it makes the program messy but because it does suggest a duplication of very heavy computation (here we suppose that the program `factorial(100)` calculates the factorial of 100, which in many cases results in an overflow). The former reason may be very important from the view of software engineering where readability and reusability of programs are essential, but it is not a matter to be discussed now. Here we shall stick to the second point - efficiency. Many programmers should agree to rewrite the expression above as

```
let x = factorial(100) in ... (x + 123) * x ...
```

The intention is that, we avoid calculating `factorial(100)` twice by *sharing* the result of this computation, without changing the result of computation. The `let` syntax indicates that `factorial(100)` is a shared resource with a name x which are later referred (used) at two places in the program.

But actually this is not just a matter for programmers, but more essentially the problem of the implementor of the programming language. Though the two examples above are supposed to return the same result, hence are extensionally equivalent, they are "intensionally" different because the amount of the involved computation is different; implementors must realize some semantic models in which such these two have distinct denotations – they may not be models for programmers (who just care about the results) but are models for implementors (who care about the actual computational steps behind the results).

Graph rewriting theory – the theory of *sharing graphs* (*term graphs*) and rewriting systems on them – has been recognized as a canonical and practically useful instance of such models for implementors [15, 84]. The idea is to use graphs for representing the sharing relations of resources and realize computation on them as rewriting systems. For instance, the situation above can be explained simply by the graphical representation of the expressions, as

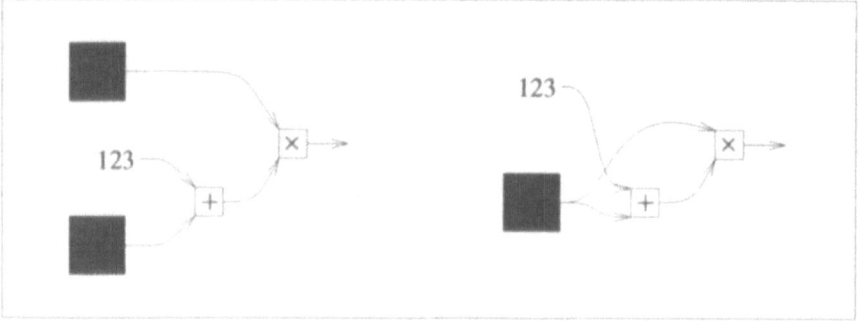

The left tree corresponds to the original unshared version, whereas the right graph is for the "refined" version with sharing of a resource. The actual computation is modeled by rewriting, i.e. local replacement of subgraphs. Obviously the left one requires more computation (rewriting steps) because of the duplicated resource (subgraph).

Impure Cases: Sharing as a Programming Technique

Consider a language with a non-deterministic construct `zero_or_one` which returns 0 or 1 at random. As before, we shall use the let-syntax for representing sharing. Then the following two programs obviously have different meanings.

```
zero_or_one + zero_or_one

let x = zero_or_one in x + x
```

The former returns 0, 1 or 2, whereas the latter 0 or 2 (see Figure 1.1). In this case the shared resource is not *pure*; it contains a side-effect, thus should be better understood as a process in a concurrent language or an object in an object-oriented language. Similar things happen if we consider imperative languages with states. In such "impure" settings, introducing sharing may change the result of computation, hence changing the extensional (programmers') semantics of the language. Therefore sharing becomes an important feature of the programming language which programmers have to recognize as a programming technique; and actually most programmers of impure languages do, often explicitly when manipulating states, objects and memories.

Cyclic Sharing and Recursion

Circular phenomena have been a rich source of a wide range of intellectual investigations for long time – in science, technology, and even philosophy; see [16] for a survey and lots of examples. Computer science is not an exception. Sharing graph-based models have a natural advantage in representing cyclic data structures, and the most interesting and practical usage of such cyclic sharing is, of course, as the means of realizing recursive computation, which is one of the most important subjects in computer science. As already shown by Turner [87] in 70s, recursive computation can be efficiently implemented using self-referential (i.e. cyclic) terms. We come back this point later and explain in some detail – the analysis of recursive computation created from cyclic sharing is one of the central implications of this thesis.

1.3 Sharing Graphs and Their Presentation

As motivated above, we regard sharing graphs, or term graphs, as abstract representations of the sharing relation of resources. They can be seen as a special sort of directed graph in which nodes represent resources and links show the sharing, but perhaps are better understood as a generalization of the tree notations for terms – the name "term graphs" means the direct generalization of "term trees".

If there is no notion of sharing, it suffices to talk about just trees (terms) where subtrees (subterms) correspond to subcomputations. However, if we want to talk about sharing, trees are not sufficient, and we are naturally led to replace trees by a class of directed graphs. Now a subgraph may be referred from various places in the graph,

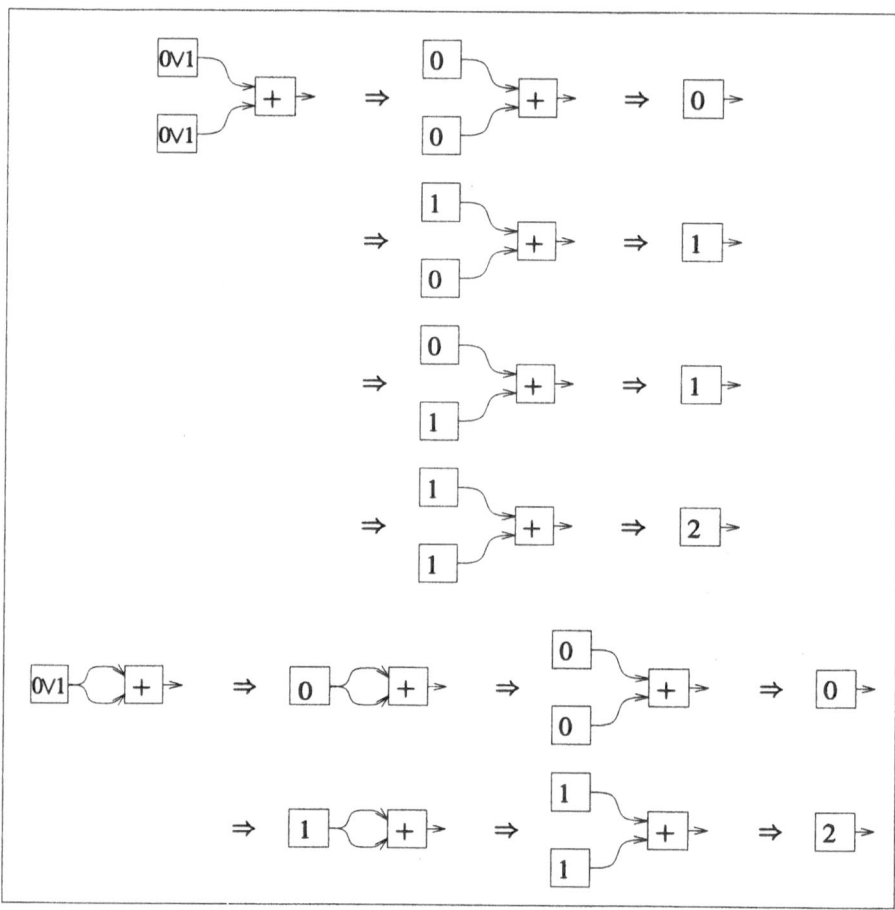

Figure 1.1: Sharing non-deterministic computation

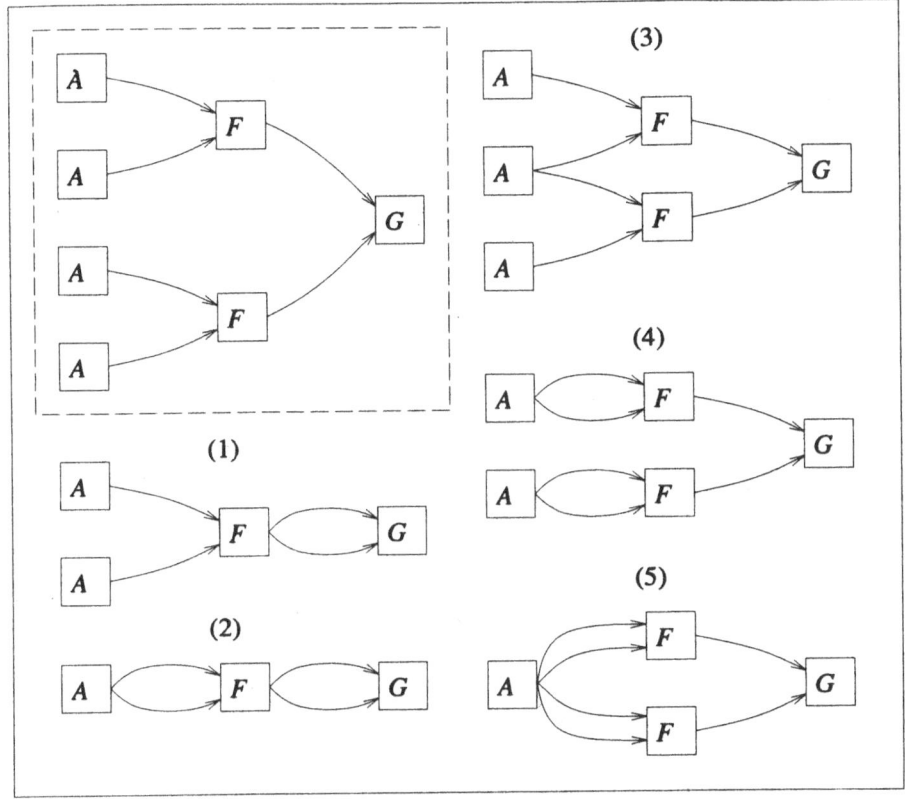

Figure 1.2: Various term graphs corresponding to a term

thus representing a shared resource. Figure 1.2 shows that there are various sharing graphs corresponding to a term $G(F(A,A),F(A,A))$. As mentioned earlier, the meaning of sharing changes depending on the computation concerned. If each node represents purely functional computation, the difference between these sharing graphs is just about the amount of computation. The final answer will be the same, but the sharing graph (2) presents the optimal way to get the answer. On the other hand, if A is a process which returns 0 or 1 non-deterministically and F and G calculate the sum of arguments, then the original term presents a computation which returns 0, 1, 2, 3 or 4, while (1) and (4) return 0, 2 or 4, whereas (2) and (5) returns just 0 or 2. (3) returns 0, 1, 2, 3 or 4 as the original term, but the probability would be changed.

Allowing cyclic bindings, sharing graphs get further flexibility. Let us look at some instances of cyclic sharing graphs (Figure 1.3). (1) and (2) present the simplest situations of cyclic sharing. In (1), the resource I refers to itself; (2) may seem odd as it does not involve any resource, but such a "self-referential pointer" or "trivial cycle" can occur even in a realistic situation. (3) is similar to (1), except that it has one additional input. A more sophisticated example is (4) where F and G mutually refer

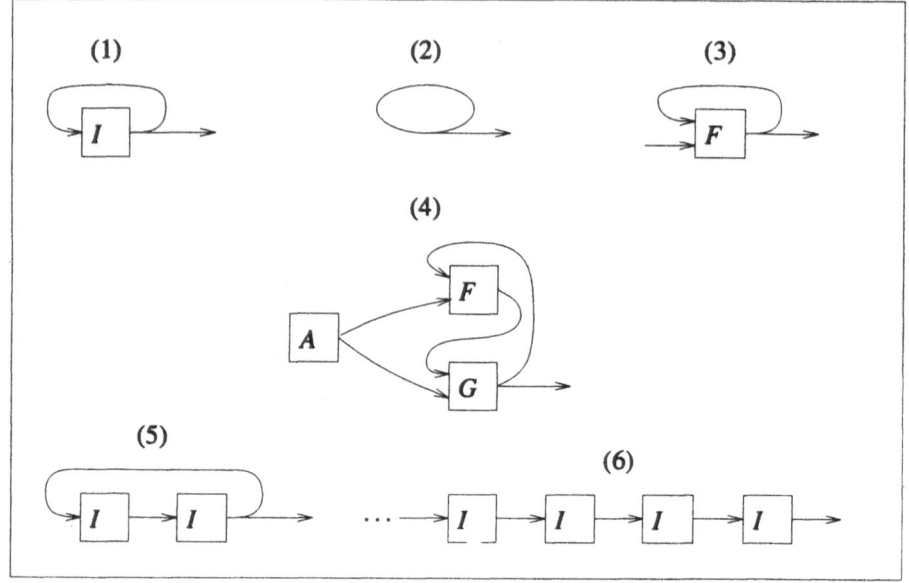

Figure 1.3: Cyclic sharing graphs

each other. (1) and (5) have the same tree-unwinding $I(I(I(I(\ldots))))$ as (6), but again it depends on the situation whether we should identify the meaning of (1) and (5) with (6).

Now we turn our attention to how to present term graphs concisely. Defining them as directed graphs, as we will do later, is not very informative; sharing graphs have more structural and algebraic properties than general directed graphs do, and we wish to capture this nature. A first hint comes from the observation above that sharing graphs can be obtained by enriching traditional terms (trees) with constructs for acyclic or cyclic sharing. Our programming example already suggests a convenient syntax for them - the let (letrec) blocks.

Actually similar notions have appeared in many places for presenting similar kind of (possibly circular) dependency relations. There are various versions of *systems of equations* for describing "non-well-founded sets" [5, 16] like

$$\begin{aligned} x &= \{y\} \\ y &= \{x,z\} \\ z &= \{x\} \end{aligned}$$

(The anti-foundation axiom states that this kind of system has a unique solution.) Similarly it is common to present a state transition system like finite state automata, and also concurrent processes, e.g. [63], by a system of equation

$$Clock \quad = \quad tick.Clock + break.Stuckclock$$

Yet another popular instance is the description of inductive (or recursive) types: for instance the type T of finite branching finite trees can be represented as a solution of a system of equations

$$T = F$$
$$F = 1 + T \times F$$

(The terms can be generated by BNF

$$t ::= \text{span}(f)$$
$$f ::= \text{nil} \mid \text{cons}(t, f) .)$$

These systems of equations have natural graph presentations, though it is possible that two different systems may describe the identical graph[1]. So there should be an equational theory on these systems which is sound and complete with respect to the graph interpretation.

We give such an axiomatization on our terms with the let/letrec blocks (which are of course an instance of systems of equations). Such notation has an advantage in allowing us equational and inductive structural reasoning about sharing graphs. We inductively construct (the presentations of) sharing graphs from variables (pointer names), function symbols (resources) and systems of equations. Thus, as the traditional algebraic theories for terms, we give equational theories for sharing graphs in terms of systems of equations for which we use the let/letrec-binding syntax. For instance, the acyclic sharing graphs in the first example can be presented as

(1)   let $x = F(A, A)$ in $G(x, x)$
(2)   let $y = A$ in let $x = F(y, y)$ in $G(x, x)$
(3)   let $y = A$ in $G(F(A, y), F(y, A))$
(4)   let $y = A$ in let $y' = A$ in in $G(F(y, y), F(y', y'))$
(5)   let $y = A$ in $G(F(y, y), F(y, y))$

As noted above, two different terms can represent the same graph; for instance, (3) can be presented as let $y = A$ in let $x = F(A, y)$ in $G(x, F(y, A))$, and our equational theory guarantees that this is equal to let $y = A$ in $G(F(A, y), F(y, A))$. Similarly, the (finite) cyclic sharing graphs in the second picture correspond to

(1)   letrec $x = I(x)$ in x
(2)   letrec $x = x$ in x
(3)   letrec $x = F(y, x)$ in x
      (the free variable y represents the unspecified input node)
(4)   letrec $x = A$, $y = F(x, z)$, $z = G(x, y)$ in z
(5)   letrec $x = I(I(x))$ in x

A simple discipline of typing is naturally given, as for traditional algebraic theories, in which any sharing graph is equipped with its input and output types (sorts). This allows us to construct graphs by well-typed composition inductively.

---

[1] Actually, for these examples, we usually work up to some stronger equivalences than that of graphs; for instance two systems are often equated if they correspond to the same infinite unwinding, equivalently if they are "bisimular". But here we do not presuppose such specific semantic interpretations, and just compare the graphs concerned themselves.

Moreover, the rewriting rules on sharing graphs are easily presented on such an equational formulation, in similar manner to the usual term rewriting rules on algebraic theories. The only difference is that in each rewriting step we replace a subgraph by another (with the same typing), instead of replacing a subterm by another.

Such advantages of this style of presentation have already been emphasized and studied by Klop, Ariola and others in the context of graph rewriting theory [7, 11]. In this thesis we basically follow their ideas, but use them freely in a more general and semantic (algebraic) context. The merit of the equational presentation becomes clearer in developing the semantic counterpart of sharing graphs, as explained below.

1.4   Categorical Models for Sharing Graphs

Traditionally, the semantic account of sharing graphs has been given in specific models, most importantly as tree unwindings where two sharing graphs are identified if they represent the same (possibly infinite) tree. Such a semantics stands out if we use sharing graphs for representing efficient implementations of pure functional computation. In this thesis, however, we take a different starting point, for the following reasons.

1. We wish to keep as many choices of semantic models as possible, so that we can interpret various (impure) forms of computation flexibly. For instance, if we want to take non-determinism into account, the infinite tree unwinding semantics is inconsistent. Rather than starting from specific models and trying to interpret actual computation in them, we axiomatize the properties needed by the models of sharing, and then find intended models.

2. We wish to talk about the *class* of models. This enables us to prove general results on all models at once, and also to classify models in a natural manner. For instance, we will give relations between our sharing graphs and intuitionistic linear logic by comparing the classes of models.

For describing the classes of models of sharing graphs, we find category-theoretical languages useful. The canonical examples of the use of category theory in this direction are the correspondence between algebraic theories and cartesian categories (categories with finite products), as well as that between the simply typed lambda calculus and cartesian closed categories. Let us summarize these "standard" *categorical type theory* correspondence as below; to make the connection with cyclic sharing, we include the treatment of recursion in our picture (Figure 1.4). Following Lawvere [58], we give models of an algebraic theory by a *finite product preserving functor* from the *classifying category* (term model) of the algebraic theory into a cartesian category. Each function symbol F with arity $((\sigma_1, \ldots, \sigma_n), \tau)$ is interpreted as an morphism $[\![F]\!] : [\![\sigma_1]\!] \times \ldots \times [\![\sigma_n]\!] \to [\![\tau]\!]$ in the target cartesian category, where $[\![\sigma_i]\!]$, $[\![\tau]\!]$ present the objects associated with each sort σ_i, τ in the algebraic theory, and $\times$ is the (chosen) cartesian product. The interpretation is then inductively extended to all expressions (terms) in the algebraic theory – it determines a finite product preserving functor from the classifying category into the model category if and only if it satisfies the soundness property: if two expressions are provably equal in the theory, then

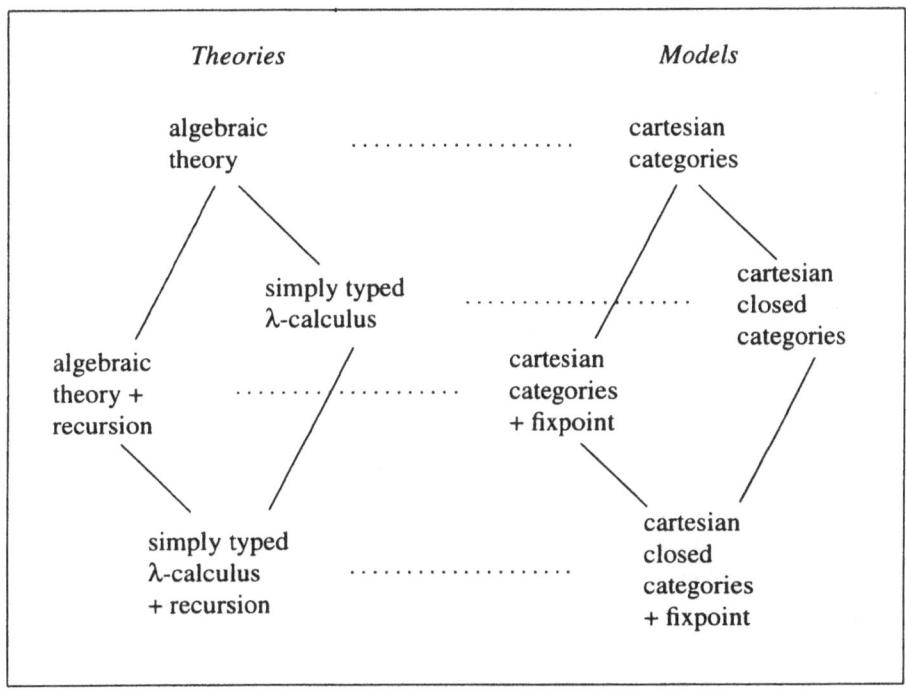

Figure 1.4: Algebraic theories and their models

their interpretation in the model is the same morphism. This is the basic picture of the theory-model correspondence in categorical type theory. A detailed account can be found, for instance, in [26].

This basic setting can be enriched with higher-order features, as well as recursive computation. For shifting to the higher-order extension, we require the existence of exponents, thus assume that the functor $(-) \times X$ has a right adjoint for each object X in the model category. Therefore we are led to the notion of *cartesian closed categories*, and again we get the theory-model correspondence between simply typed lambda theories and cartesian closed categories [26, 56] (this time the semantic interpretations are given as cartesian closed functors).

For recursion, the standard way is to assume a construction on the model cartesian (closed) category, called a *(parameterized) fixed point operator*

$$\frac{f : A \times X \to X}{f^\dagger : A \to X}$$

which is subject to the condition that $\langle id_A, f^\dagger \rangle; f = f^\dagger$ (to be more precise, we assume that this construction is natural in A, so that the model is sound for the interpretation of substitutions). In the standard notation for a recursion operator on algebraic theories, this corresponds to

$$\frac{\Gamma, x : \sigma \vdash M : \sigma}{\Gamma \vdash \mu x.M : \sigma}$$

with the fix point equation $\mu x.M = M[\mu x.M/x]$. Many concrete examples of such categories are found in domain theory, where cartesian closedness and existence of fixed point operators are fundamental requirements for giving the denotational semantics of programming languages.

The main technical development in this thesis is to give, for sharing graphs, a precise analog of this standard categorical type theory. The equational theory presentation of sharing graphs via the let (letrec)-syntax is already very close to the standard algebraic theories, and it is natural to expect that there is a similar theory-model correspondence for sharing graphs.

The essential change is that, instead of cartesian categories, we take identity-on-objects, strict symmetric monoidal functors from cartesian categories to symmetric monoidal categories as the basic setting for interpreting the sharing graphs. Intuitively, the domain cartesian category is used for modeling the non-linear nature of sharing graphs – pointer names, or links, and also copyable-values (if they exist), are duplicated or discarded freely, hence will be interpreted in the cartesian category as we do for algebraic theories. On the other hand, the codomain symmetric monoidal category is for interpreting linear entities in sharing graphs; since we do not duplicate or discard the shared resources which are expensive or contain some side effect, they must be treated linearly. (The reader familiar with linear logic [36] may informally understand this by the analogy with the logical connectives & and $\otimes$ of linear logic; later we will give the precise connection between our models of sharing and those of propositional intuitionistic linear logic.) The strict functor between them is to relate these non-linear and linear natures. In short, the essence of models of sharing lies in the separation of non-linear and linear features which live at the same time in the

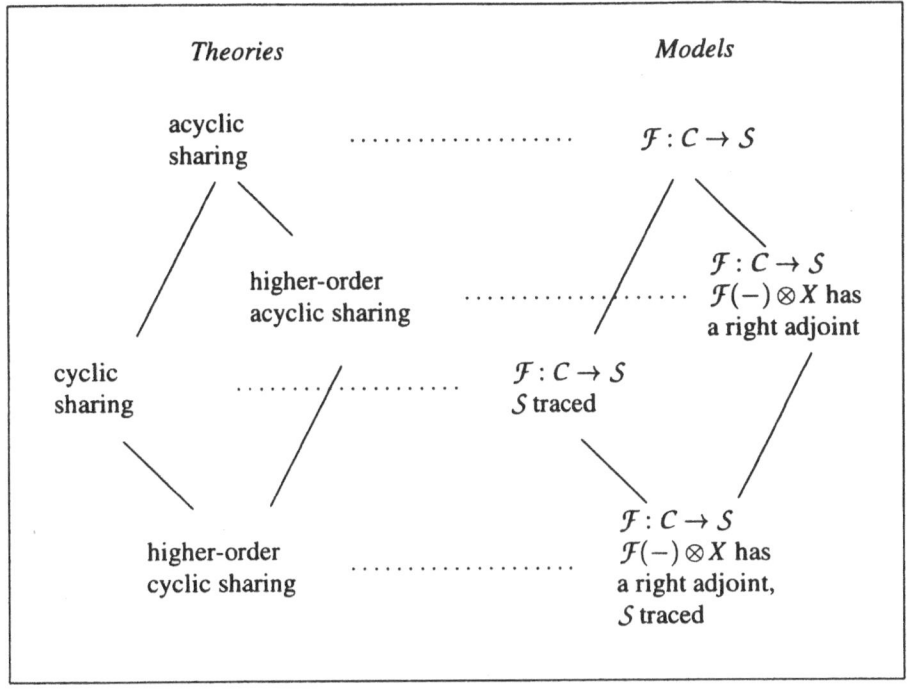

Figure 1.5: Sharing theories and their models

notion of sharing. Now we shall give our picture of the theory-model correspondence for sharing graphs (Figure 1.5). By $\mathcal{F} : C \to S$, we mean an identity-on-objects strict symmetric monoidal functor $\mathcal{F}$ from a cartesian category C to a symmetric monoidal category S.

For interpreting higher-order features, we additionally require that $\mathcal{F}(-) \otimes X$ has a right adjoint for each object X; this is the precise analog of cartesian closed categories for our setting. For interpreting cyclic sharing, we need a relatively new concept from category theory – *traced monoidal categories* [50]. Intuitively, a traced symmetric monoidal category is a symmetric monoidal category equipped with a construct for "feedback", called a *trace*:

$$\frac{f : A \otimes X \to B \otimes X}{Tr_{A,B}^{X}(f) : A \to B}$$

It would be helpful to understand that, in $Tr^{X}(f)$, f's output X is feedbacked, or linked, to f's input X. The formal axiomatization for a trace will be recalled later; we will see that it precisely corresponds to the equivalence on cyclic graphs, and the theory-model correspondence will be extended to the cyclic settings comfortably by assuming that the symmetric monoidal category S is traced.

The rewriting theories on sharing graphs are then simply modeled by the local-preorders on the symmetric monoidal category S of our models. Some graph rewriting systems, especially the *equational term graph rewriting* by Klop and Ariola, are close to our theories and their semantic models.

Note that if we restrict our attention to the case that C and S are the same cartesian category and $\mathcal{F}$ is the identity functor, then we recover the standard categorical type theory as sketched before (a connection between traces and fixed point operators will be established later).

1.5   Relating Models

To demonstrate the advantage of our generic approach, we shall relate some known systems and ours by comparing their classes of models. Many people have pointed out that term graphs have some similarity with Girard's *linear logic* [36], in their resource-sensitive natures. Also it has been pointed that Moggi's computational lambda calculus [71] looks like higher-order graph rewriting systems. We give some formal accounts to these intuitive understandings, by first relating the classes of models, and then relating the theories as a corollary.

A model of propositional intuitionistic linear logic may be described as a symmetric monoidal adjunction between a cartesian closed category and a symmetric monoidal closed category [12, 18, 20]. It is easily seen that such a structure is essentially a special case of the structures we have for interpreting acyclic sharing graphs, as sketched above. Thus there is a sound translation from the equational theory of sharing graphs into that of intuitionistic linear type theory. But we can say more: this translation is conservative, thus a linear type theory is seen as a conservative extension of the theory of sharing graphs. To prove this, we use the standard model construction technique from category theory (Yoneda construction as the free symmetric monoidal cocompletion [44]).

The connection with Moggi's work [71, 72] is much more straightforward. The models for acyclic higher-order sharing will be shown to be essentially the same as his models for computational lambda calculus, with an assumption that the associated monad has a commutative strength. As a special instance of the theory developed by Power and Robinson [77, 76], we describe this comparison.

1.6   Recursion from Cyclic Sharing

One of the traditional methods of interpreting a recursive program in a semantic domain is to use the least fixed-point of continuous functions. However, as already mentioned, in the real implementations of programming languages, we often use some kind of shared cyclic structure for expressing recursive environments efficiently. For instance, the following is a call-by-name operational semantics of the recursive call, in which free x may appear in M and N. We write $E \vdash M \Downarrow V$ for saying that evaluating a program M under an environment E results a value V; in call-by-name strategy an environment assigns a free variable to a pair consisting of an environment and a program.

$$\frac{E' \vdash N \Downarrow V \quad \text{where } E' = E \cup \{x \mapsto (E',M)\}}{E \vdash \text{letrec } x = M \text{ in } N \Downarrow V}$$

That is, evaluating a recursive program letrec $x = M$ in N under an environment E amounts to evaluating the subprogram N under a cyclic environment E' which references itself. One may see that it is reasonable and efficient to implement the recursive (self-referential) environment E' as a cyclic data structure as below.

Also it is known that if we implement a programming language using the technique of sharing, the use of the fixed point combinator causes some unexpected duplication of resources [9, 57]; it is more efficient to get recursion by cycles than by the fixed point combinator in such a setting. This fact suggests that there is a gap between the traditional approach based on fixed points and cyclically created recursion.

Our semantic models for higher-order cyclic sharing turn out to be the setting for studying recursive computation created by such a cyclic data structure, more specifically cyclic lambda graphs [10, 8]. We claim that our new models are natural objects for the study of recursive computation because they unify several aspects on recursion in just one semantic framework. The leading examples are

- the *graphical (syntactical) interpretation* of recursive programs by cyclic data structures motivated as above,

- the *domain-theoretic interpretation* in which the meaning of a recursive program letrec $x = F(x)$ in x is given by the least fixed point $\bigcup_n F^n(\perp)$, and

- the *non-deterministic interpretation* where the program letrec $x = F(x)$ in x is interpreted by $\{x \mid x = F(x)\}$, the set of all possible solutions of the equation $x = F(x)$.

Each of them has its own strong tradition in computer science. However, to our knowledge, this is the first attempt to give a uniform account on these well-known, but less-related, interpretations. Moreover, our higher-order cyclic sharing theories and cyclic lambda calculi serve as a uniform language for them.

1.7 Action Calculi as Graph Rewriting

Finally we show that our framework can accommodate Milner's *action calculi* [68], a proposed framework for general interactive computation, by showing that our sharing theories, enriched with suitable constructs for interpreting parameterized constants called controls, are equivalent to the closed fragments of action calculi [34, 75] and their higher-order/reflexive extensions [66, 67, 61].

The dynamics, the computational counterpart of action calculi, is then understood as rewriting systems on sharing theories, and interpreted as local preorders on our models. In this sense, we understand action calculi as generalized graph rewriting systems – and regard the notion of sharing as one of the fundamental concepts of action calculi.

To demonstrate how sharing is used in action calculi, we shall consider two situations representable in the action calculus-version of the π-calculus [69, 64] as presented in [68] (see Chapter 8, Example 8.1.6).

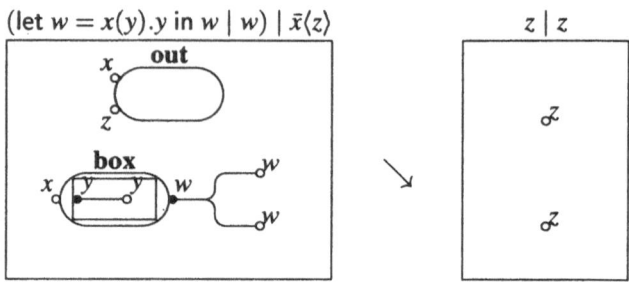

We may regard this situation (not representable in the original π-caclulus!) as a broadcasting; there is an announcer $x(y).y$ who gets a message via a telephone number x and then broadcasts it; her/his program is monitored by two listeners $w|w$. Therefore the received message z is broadcast (duplicated) to the listeners. Compare this and the unshared version $x(y).y \mid x(y).y \mid \bar{x}\langle z\rangle$, where we have two persons who share the same telephone number x. So we don't know which person will receive the message z, and there are two possible reactions (in both cases the result is $x(y).y \mid z$, thus one person remains unchanged:

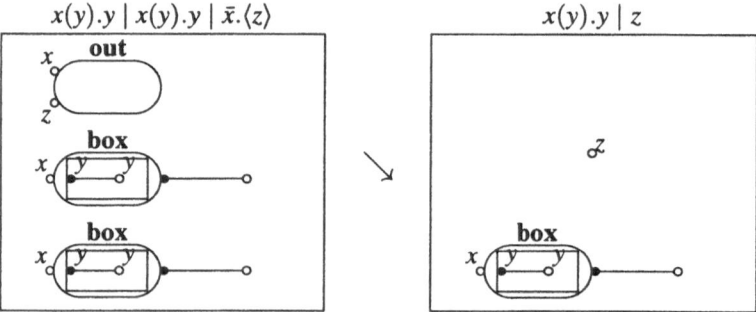

Further sophisticated and complicated examples will be available by allowing cyclic sharing (reflexion) and higher-order constructions.

All of our semantic results on sharing graphs equally apply to action calculi (with some care on the treatment of controls). The conservativity of intuitionistic linear type theory over action calculi (as reported in [13]), the correspondence between higher-order action calculi and Moggi's work (as described in [35]), and the analysis of recursive computation in reflexive action calculi (c.f. [61]) are obtained as corollaries of results on sharing graphs.

Figure 1.6 is a summary of the correspondence between our theory of sharing graphs and action calculi:

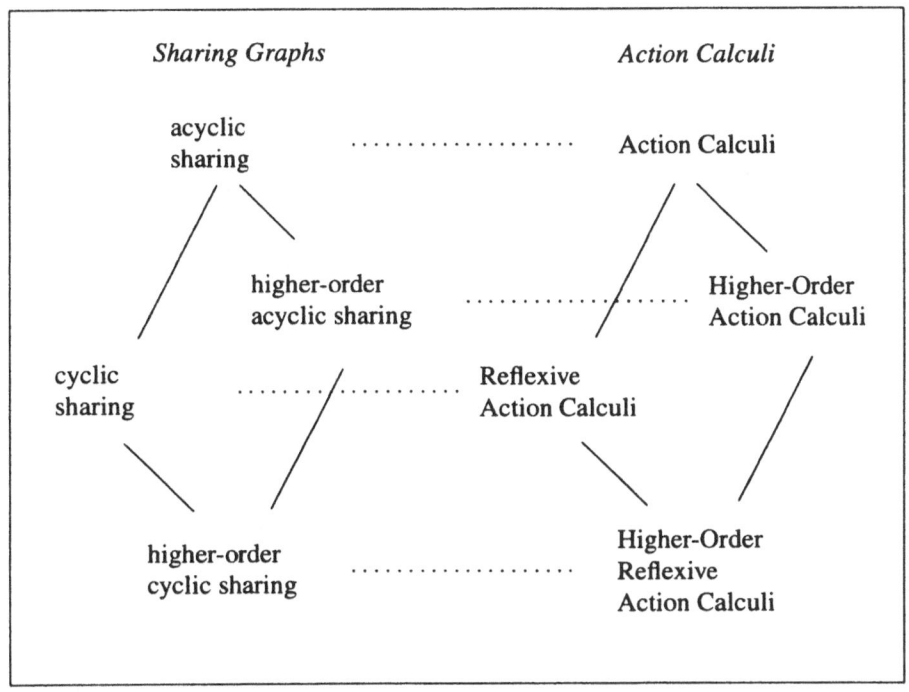

Figure 1.6: Sharing theories and action calculi

We hope that our work provides a bridge between graph rewriting theory and concurrency theory.

1.8   Overview

Chapter 2 introduces the notion of sharing graphs and the corresponding simply typed equational theories, called sharing theories. We emphasize the algebraic, structural nature of sharing graphs via the equational presentations, which leads us to the semantic development in the following chapters.

In Chapter 3 we study the category-theoretic models of acyclic sharing theories. In terms of symmetric monoidal categories and functors, we describe the class of models, and establish the soundness and completeness, in a similar way to the standard categorical type theory.

In Chapter 4 we give a higher-order extension of acyclic sharing. The models of this setting are obtained by assuming additional conditions formulated as adjunctions, and we repeat the same pattern as in Chapter 3.

As an application of our approach, in Chapter 5 we relate our acyclic sharing theories with notions of computation and intuitionistic linear type theory by comparing their classes of models.

In Chapter 6 we give the models of cyclic sharing, by additionally using the notion of traced monoidal categories. After reviewing traced monoidal categories, we establish the expected properties of our models, again in the same way as Chapter 3.

Chapter 7 describes higher-order cyclic sharing. The models of this setting, obtained by combining those in Chapter 4 and Chapter 6, are of particular interest as they support a generalized form of recursive computation. We look at this in some detail, together with the connection with cyclic lambda calculi.

Chapter 8 is devoted to showing that Milner's action calculi can be accommodated in our framework.

Finally, in Chapter 9, we conclude this thesis with some discussions towards further research.

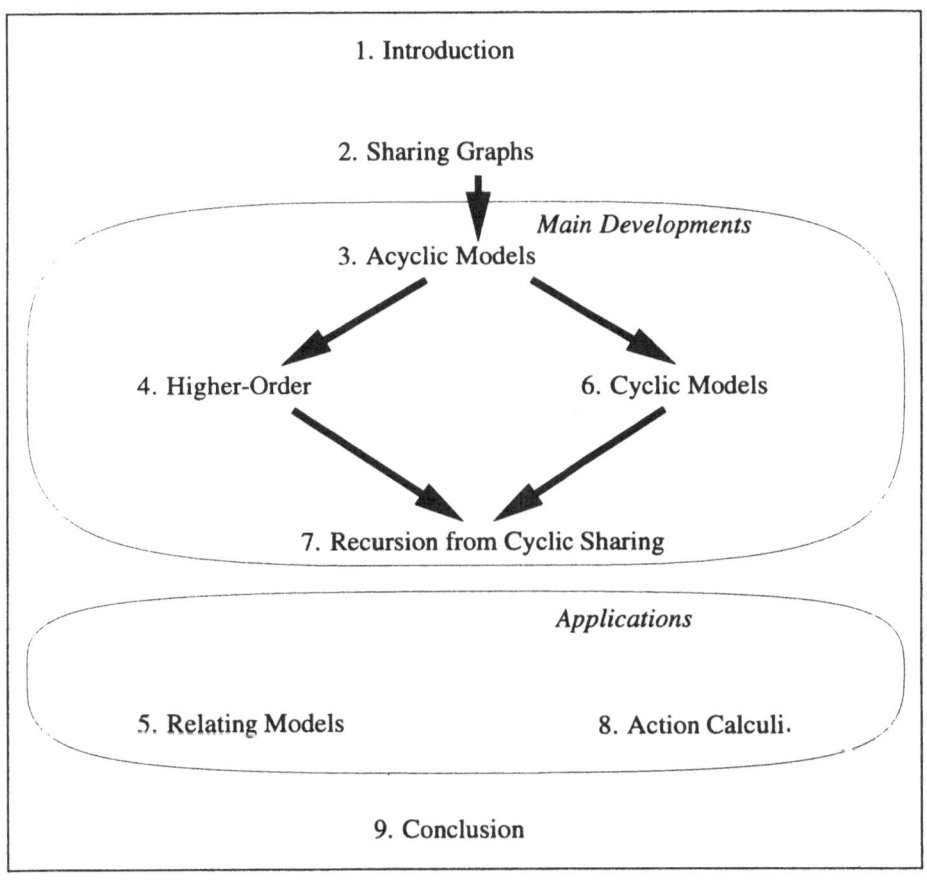

Figure 1.7: Overview of this book

2
Sharing Graphs and Equational Presentation

Following the standard way of describing graphical structures, we first formulate sharing graphs as directed graphs with extra information and conditions, as found in the literature (though we give a mildly generalized version in which multiple roots, or conclusions, are permitted).

However, such a description is always lengthy, and technically not easy to manipulate. Also the algebraic, structural nature of sharing graphs is not clear in such formulations. Following an old idea of representing sharing graphs as systems of equations, we give an equational, in other words algebraic, presentation of finite sharing graphs.

In the same way that trees are represented as term expressions in algebraic theories, our sharing graphs are represented as term expressions of mildly relaxed algebraic theories enriched with constructs for sharing (let/letrec bindings). We establish the desired equivalence between the graph-theoretical description and the equational presentation of finite sharing graphs, for both acyclic and cyclic cases.

The results in this chapter are conceptually and technically not new at all – similar ideas have been around for long time. However, our intention here is to use the equational presentation of sharing graphs for emphasizing the algebraic aspects of them, which later naturally lead us to the formulation of the semantic models of sharing graphs. Therefore this chapter should be read as a preparation for our main technical developments.

2.1 Sharing Graphs

We first fix the notion of a signature, on which our sharing graphs are constructed:

Definition 2.1.1 (signature)
Let S be a set of sorts. An (finitary) *S-sorted signature* is a set Σ of operation symbols together with an arity function assigning to each operation symbol F a pair of finite lists $((\sigma_1, \ldots, \sigma_m), (\tau_1, \ldots, \tau_n))$ of S's elements. Notation:

$$F : (\sigma_1, \ldots, \sigma_m) \longrightarrow (\tau_1, \ldots, \tau_n).$$

$\square$

Remark 2.1.2 The main difference between our definition of signatures and the more traditional one is that we allow multiple conclusions of an operator symbol. Intuitively, we allow operators which return tuples of results; or even operators which do not return anything (which, however, does not mean that such operators do not have computational significance – in some impure settings, they may have some side effects!). If the reader is familiar with algebraic theories with cartesian product types, this formulation could be considered as a mild variant, though our multiple conclusions will not form cartesian products, but symmetric monoidal products. So, while in the standard algebraic theories a term of product types can be regarded as a tuple of terms, this is not the case for our sharing graphs, where an "indecomposable" resource can have multiple outputs. □

For $F : (\sigma_1, \ldots, \sigma_m) \longrightarrow (\tau_1, \ldots, \tau_n)$, we may write $\mathrm{dom}(F)$ for $(\sigma_1, \ldots, \sigma_m)$, $\mathrm{cod}(F)$ for $(\tau_1, \ldots, \tau_n)$, $\mathrm{dom}(F)_i$ for σ_i (*i*th input sort) and $\mathrm{cod}(F)_j$ for τ_j (*j*th output sort).

In formulating (cyclic) sharing graphs, we need care about *trivial cycles* (also known as "blackholes") as repeatedly pointed in the literature of graph rewriting theory (in connection with the "cyclic-I problem", which will be discussed in Example 2.4.3), see e.g. [11]. A trivial cycle is a pointer which does not refer any resource but *itself* – just like a program "letrec x be x in x" which does not involve any computational resource but represents a circularly bound pointer. For both practical and technical reasons, we do not want to exclude such trivial cycles from our sharing graphs. To accommodate them, we need an additional constant for each sort:

Definition 2.1.3 (signature with •)
Given an S-sorted signature Σ, we define an S-sorted signature $\Sigma_{\bullet}$ by additionally assuming an operation symbol $\bullet^{\sigma} : () \to (\sigma)$ for each sort σ. □

A rooted directed graph (with a label on each node) is specified by a set of nodes V, a labeling function L from V to the set of labels and a set of edges $E \subseteq V \times V$ together with a specified root node $c \in V$. For describing our sharing graphs, we need more information as follows. First, an operation symbol may have multiple inputs and outputs, so we need to specify which input is linked to which output, in terms of an "argument function" A. Second, since we allow multiple outputs of the graph, we have to specify a list of outputs $(c_1, \ldots, c_n)$, rather than a single root node c. Thus we have a tuple $(V, L, A, (c_1, \ldots, c_n))$. Moreover, we want to make the graph well-typed, that is, an input and an output can be linked only when they have the same sort. So we assume a constraint for ensuring the well-typedness. Formally:

Definition 2.1.4 (sharing graph)
We fix a countable set $\{d_1, d_2, \ldots\}$. A (finitary) *sharing graph* over an S-sorted signature Σ of type $(\sigma_1, \ldots, \sigma_m) \to (\tau_1, \ldots, \tau_n)$ is a tuple $(V, L, A, (c_1, \ldots, c_n))$ such that

- V is a set.

- L is a function from V to $\Sigma_{\bullet}$.

- A is a function from V to C^*, such that $|A(v)| = |\mathrm{dom}(L(v))|$, where

$$C = \{\langle v, j \rangle \mid v \in V, 1 \leq j \leq |\mathrm{cod}(v)|\} \oplus \{d_1, \ldots, d_m\}.$$

Write $A(v)_i$ for the ith component of $A(v)$.[1] C serves as the set of all outputs (codomains), while d_i is just a name for the i-th input (domain).[2]

- $c_i \in C$ for $1 \leq i \leq n$.

- Condition on types:

  - For $1 \leq i \leq |A(v)|$, $\text{dom}(L(v))_i = \begin{cases} \text{cod}(L(w))_j & \text{if } A_i(v) = \langle w, j \rangle \\ \sigma_j & \text{if } A_i(v) = d_j \end{cases}$

  - If $c_i = \langle v, j \rangle$, $\text{cod}(L(v))_j = \tau_i$. If $c_i = d_j$, $\sigma_j = \tau_i$.

  □

Example 2.1.5 Consider a sort $S = \{\text{nat}\}$ and a signature

$$\Sigma = \{\text{zero} : () \to (\text{nat}), \ \text{plus} : (\text{nat},\text{nat}) \to (\text{nat})\}.$$

Let us construct sharing graphs of type $() \to (\text{nat})$ as drawn in the pictures below.

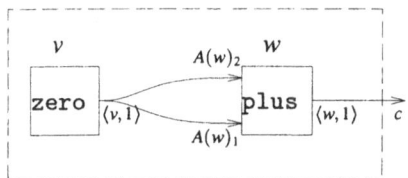 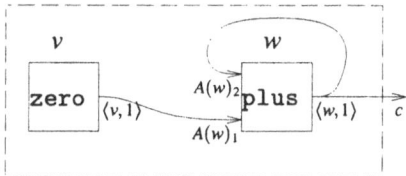

A sharing graph $(V, L, A, (c)) : () \to (\text{nat})$ for the left picture may be specified by $V = \{v, w\}$, $L(v) = \text{zero}$, $L(w) = \text{plus}$, $A(w)_1 = A(w)_2 = \langle v, 1 \rangle$, and $c = \langle w, 1 \rangle$. For the right one, V, L and c are unchanged but we modify A as $A(w)_1 = \langle v, 1 \rangle$ and $A(w)_2 = \langle w, 1 \rangle$. □

Example 2.1.6 If V is empty, A and L are the unique functions from the empty set, while $C = \{d_1, \ldots, d_m\}$. Thus such a sharing graph is determined by a function from $\{1, \ldots n\}$ to $\{1, \ldots, m\}$ subject to the type constraint. Below are two examples of such sharing graphs of type $(\tau, \sigma, \tau) \to (\sigma, \tau, \tau)$.

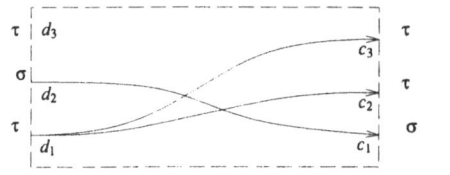 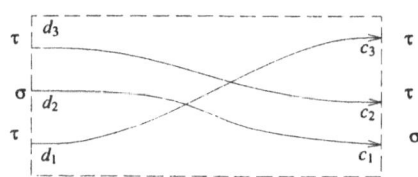

The left graph is specified by output nodes $c_1 = d_2$, $c_2 = d_1$ and $c_3 = d_1$. Similarly, the right one is specified as $c_1 = d_2$, $c_2 = d_1$ and $c_3 = d_1$. □

---

[1] For sets S and S', we write S^* for the set of finite lists of elements of S, and $S \oplus S'$ for the disjoint union of S and S'.

[2] Instead of $\{d_1, \ldots, d_m\}$, we can simply use natural numbers $\{1, \ldots, m\}$ – we did not do so just for the readability.

Example 2.1.7 A more tangled example. For

$$S = \{\texttt{bool},\texttt{nat}\} \text{ and } \Sigma = \{\texttt{F} : (\texttt{bool},\texttt{nat}) \to (\texttt{nat},\texttt{bool})\},$$

a sharing graph $(V, L, A, (c_1, c_2, c_3))$ of type $(\texttt{bool},\texttt{nat}) \to (\texttt{nat},\texttt{nat},\texttt{bool})$ may be given as below.

$V = \{v_1, v_2, v_3, v_4\}$.
$L(v_1) = L(v_2) = L(v_3) = F, L(v_4) = \bullet$.
$A(v_1)_1 = \langle v_1, 2 \rangle$, $A(v_1)_2 = d_2$, $A(v_2)_1 = d_1$, $A(v_2)_2 = \langle v_3, 1 \rangle$, $A(v_3)_1 = \langle v_2, 2 \rangle$ and $A(v_3)_2 = \langle v_4, 1 \rangle$.
$c_1 = \langle v_2, 1 \rangle$, $c_2 = \langle v_3, 1 \rangle$ and $c_3 = \langle v_3, 2 \rangle$.

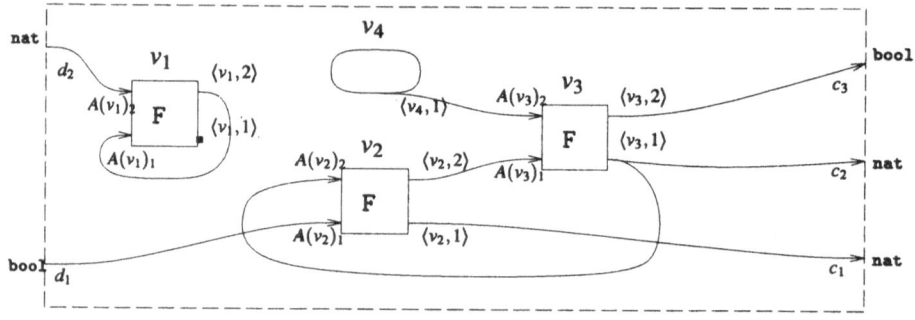

Note that v_4 ($\bullet^{\texttt{nat}}$) is drawn as a trivial cycle.   □

Remark 2.1.8 The direction of the links drawn in the pictures is opposite to that of most of the pictorial presentations of sharing graphs in the literature. The reason we invert the direction will become clear when we introduce the categorical semantics of sharing graphs: our direction is that of morphisms in our semantic categories.   □

Remark 2.1.9

- If we just consider operator symbols with just one output sort, then in the definition above C becomes $V \cup \{d_1, \dots, d_m\}$ and we recover a standard definition like in [11] (c.f. [8]).

- In a standard terminology, assuming m inputs amounts to assuming m free variables $\{d_1, \dots, d_m\}$.

- There is no technical difficulty to formulate the *infinitary* version of sharing graphs, by allowing infinitely many inputs and outputs (and allowing operators with infinitely many inputs and outputs too). However we do not see any practical benefit of such extra generality for our study (though infinitary operators must be included in some related settings, e.g. systems of equations for non-well-founded set theory [5, 16] which is out of the scope of this thesis); and giving the semantic models of the infinitary setting is far more complicated than the case for the finitary one, so we do not consider such a version. Also in this thesis we are mainly interested in the finite sharing graphs (see below), where the infinitary version seems meaningless.   □

Definition 2.1.10 (finite sharing graph)
A sharing graph $(V, L, A, (c_1, \ldots, c_n))$ is *finite* if $|V| < \infty$. □

Definition 2.1.11 (equivalence of sharing graphs)
Two sharing graphs $G = (V, L, A, (c_1, \ldots, c_n))$, $G' = (V', L', A', (c'_1, \ldots, c'_n))$ of the same type are said to be *equivalent* if there is a bijection on sets $f : V \xrightarrow{\sim} V'$ such that $L' \circ f = L$, $A' \circ f = (f \times id)^* \circ A$ and $(f \times id)(c_i) = c'_i$. Obviously this determines an equivalence relation on sharing graphs of the same type, and we write $G \equiv G'$ if G and G' are equivalent. □

For instance, it is easily seen that two sharing graphs in Example 2.1.5 are not equivalent.

Definition 2.1.12 (dependency relation)
Let $(V, L, A, (c_1, \ldots, c_n))$ be a sharing graph. We define a binary relation $<$ on V by $v < w$ if $A(w)_i = \langle v, j \rangle$ for some i, j, and also $v < v$ if $L(v) = \bullet$. Then the *dependency relation* $<^*$ is the transitive closure of $<$. □

For instance, in Example 2.1.7 the dependency relation consists of $v_1 <^* v_1$, $v_2 <^* v_2$, $v_2 <^* v_3$, $v_3 <^* v_2$, $v_3 <^* v_3$, $v_4 <^* v_3$ and $v_4 <^* v_4$.

Definition 2.1.13 (acyclic sharing graph)
A sharing graph $(V, L, A, (c_1, \ldots, c_n))$ is *acyclic* if there is no $v \in V$ such that $v <^* v$.
□

For example, one may see that the left graph in Example 2.1.5 is acyclic while the right is not (i.e. truly cyclic).

In the sequel, by a sharing graph, we may mean the equivalence class of the sharing graph (if there is no confusion). And we work just on finite sharing graphs unless explicitly mentioned.

2.2 Acyclic Sharing Theory

In this section we give the equational presentation of acyclic sharing graphs (cyclic graphs are dealt with in the following section). As discussed informally in the introduction, our presentation is based on "systems of equations", represented by the let-bindings in the acyclic case. So the expressions are terms of traditional algebraic theories plus let-blocks, and we assume a set of axioms which ensures that two expressions correspond to the same sharing graph if and only if they are provably equal in the theory. Though we need some care for dealing with multiple outputs (represented by tensor products), our development is fairly close to the standard stories for (many sorted) algebraic type theories as found in [26] and we hope that our syntax is not very far from such traditional treatments. The comparison between sharing graphs and the equivalence classes of the terms of sharing theories is done by giving the translation between them, using a normal form property of the theories.

In the rest of this chapter, we fix a set of sorts S and a signature Σ, as introduced in the last section, unless otherwise stated.

Definition 2.2.1 (raw expressions)

$$M ::= x \mid F(M) \mid 0 \mid M_1 \otimes M_2 \mid \text{let } (x_1, \ldots, x_m) \text{ be } M_1 \text{ in } M_2$$

We assume that 0 and $\otimes$ satisfy *strict associativity*: we identify $0 \otimes M$, $M \otimes 0$ with M, and $(M_1 \otimes M_2) \otimes M_3$ with $M_1 \otimes (M_2 \otimes M_3)$ (and write $M_1 \otimes M_2 \otimes M_3$ for it). □

The notion of free and bound variables is defined as usual, and we write $FV(M)$ and $BV(M)$ for the sets of free and bound variables of the expression M respectively. We write $M\{N/x\}$ for a capture-free substitution of N into free x's in M. Similarly we use $M\{N_1/x_1, N_2/x_2\}$, $M\{\vec{N}/\vec{x}\}$ etc for simultaneous substitutions. In the sequel, let $(x_1, \ldots, x_m)$ be M_1 in M_2 may be written as let $(\vec{x})$ be M_1 in M_2 for short. Also we may write $\vec{x}$ instead of $x_1 \otimes \ldots \otimes x_m$, which includes 0 as the case of $m = 0$.

Definition 2.2.2 (typing)
In the following, by a *context* we mean a finite list of pairs of variables and sorts like $x_1 : \sigma_1, \ldots, x_m : \sigma_m$ where x_i are different to each other. We say a term M has a type $(\sigma_1, \ldots, \sigma_n)$ under a context Γ if $\Gamma \vdash M : (\sigma_1, \ldots, \sigma_n)$ is derivable from the following typing rules. Such a term is called a *well-typed term*.

$$\frac{}{\Gamma, x : \sigma \vdash x : (\sigma)} \text{ variable}$$

$$\frac{\Gamma \vdash M : (\sigma_1, \ldots, \sigma_m) \quad F : (\sigma_1, \ldots, \sigma_m) \rightarrow (\tau_1, \ldots, \tau_n)}{\Gamma \vdash F(M) : (\tau_1, \ldots, \tau_n)} \text{ operator}$$

$$\frac{}{\Gamma \vdash 0 : ()} \text{ unit}$$

$$\frac{\Gamma \vdash M : (\sigma_1, \ldots, \sigma_m) \quad \Gamma \vdash N : (\tau_1, \ldots, \tau_n)}{\Gamma \vdash M \otimes N : (\sigma_1, \ldots, \sigma_m, \tau_1, \ldots, \tau_n)} \text{ tensor}$$

$$\frac{\Gamma \vdash M : (\sigma_1, \ldots, \sigma_m) \quad \Gamma, x_1 : \sigma_1, \ldots, x_m : \sigma_m \vdash N : (\tau_1, \ldots, \tau_n)}{\Gamma \vdash \text{let } (x_1, \ldots, x_m) \text{ be } M \text{ in } N : (\tau_1, \ldots, \tau_n)} \text{ let}$$

$$\frac{\Gamma, x : \sigma, x' : \sigma', \Gamma' \vdash M : (\tau_1, \ldots, \tau_n)}{\Gamma, x' : \sigma', x : \sigma, \Gamma' \vdash M : (\tau_1, \ldots, \tau_n)} \text{ exchange}$$

□

On notation: a list of sorts $(\sigma_1, \ldots, \sigma_m)$ may be abbreviated to $(\vec{\sigma})$.

Definition 2.2.3 (axioms)

(σ_{var})	let (x) be y in M	$=$	$M\{y/x\}$
(id)	let $(\vec{x})$ be M in $\vec{x}$	$=$	M
(ass_1)	let $(\vec{x})$ be (let $(\vec{y})$ be L in M) in N	$=$	let $(\vec{y})$ be L in let $(\vec{x})$ be M in N
(ass_2)	let $(\vec{x})$ be L in let $(\vec{y})$ be M in N	$=$	let $(\vec{x},\vec{y})$ be $L \otimes M$ in N
$(\otimes_1)$	$L \otimes ($let $(\vec{x})$ be M in $N)$	$=$	let $(\vec{x})$ be M in $L \otimes N$
$(\otimes_2)$	(let $(\vec{x})$ be L in $M) \otimes N$	$=$	let $(\vec{x})$ be L in $M \otimes N$
(subst)	let $(\vec{x})$ be M in $F(N)$	$=$	$F($let $(\vec{x})$ be M in $N)$

They are "equations in contexts"; For each axiom, under the same context, the left hand side must have the same type with the right hand side. □

For instance, in (ass_2), x's cannot be free in M.

Definition 2.2.4 (acyclic sharing theory)
An *acyclic sharing theory* over Σ is an equational theory on the well-typed terms closed under the term construction described above, where the equality on terms is a congruence relation containing the axioms above, i.e. equations derivable from the following inference rules, possibly with additional axioms.

$$\frac{\Gamma \vdash M : (\vec{\tau}) \quad \Gamma \vdash N : (\vec{\tau}) \quad M = N \text{ is included in the axioms}}{\Gamma \vdash M = N : (\vec{\tau})}$$

$$\frac{\Gamma \vdash M : (\vec{\tau})}{\Gamma \vdash M = M : (\vec{\tau})} \qquad \frac{\Gamma \vdash M = N : (\vec{\tau})}{\Gamma \vdash N = M : (\vec{\tau})} \qquad \frac{\Gamma \vdash L = M : (\vec{\tau}) \quad \Gamma \vdash M = N : (\vec{\tau})}{\Gamma \vdash L = N : (\vec{\tau})}$$

$$\frac{\Gamma \vdash M = N : (\vec{\sigma}) \quad F : (\vec{\sigma}) \to (\vec{\tau})}{\Gamma \vdash F(M) = F(N) : (\vec{\tau})} \qquad \frac{\Gamma \vdash M = M' : (\vec{\sigma}) \quad \Gamma \vdash N = N' : (\vec{\tau})}{\Gamma \vdash M \otimes N = M' \otimes N' : (\vec{\sigma},\vec{\tau})}$$

$$\frac{\Gamma \vdash M = M' : (\vec{\sigma}) \quad \Gamma,\vec{x}:\vec{\sigma} \vdash N = N' : (\vec{\tau})}{\Gamma \vdash \text{let } (\vec{x}) \text{ be } M \text{ in } N = \text{let } (\vec{x}) \text{ be } M' \text{ in } N' : (\vec{\tau})}$$

$$\frac{\Gamma,x:\sigma,x':\sigma',\Gamma' \vdash M = N : (\vec{\tau})}{\Gamma,x':\sigma',x:\sigma,\Gamma' \vdash M = N : (\vec{\tau})}$$

By the *pure acyclic sharing theory*, we mean the acyclic sharing theory with no additional axioms. □

In the sequel, the word "acyclic" may be dropped, unless there can be a confusion with the cyclic sharing theories which will be introduced later.

To help with the intuition, we give a pictorial account for our term constructions:

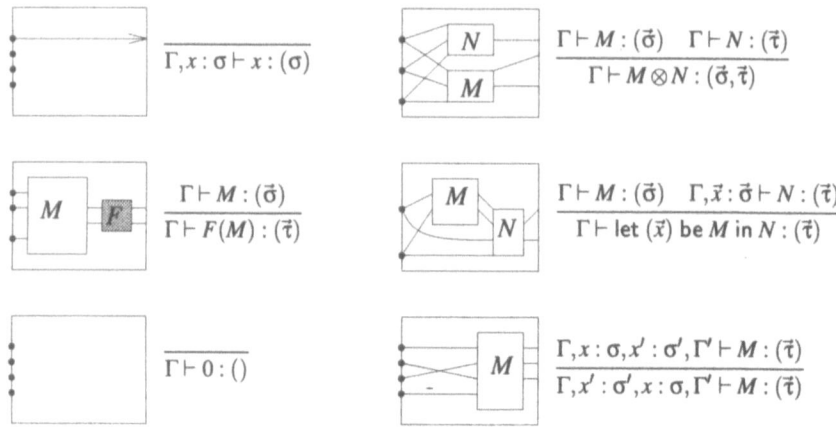

Each term construction (type derivation) amounts to constructing a new graph from existing ones.

Remark 2.2.5 It might be helpful to understand the let-bindings as "suspended (or delayed) substitutions"; they perform the real substitution only for the case of variables (σ_{var}). If we write $N[M/\vec{x}]$ for let $(\vec{x})$ be M in N, the axioms can be rewritten as

(σ_{var})	$M[y/x]$	$=$	$M\{y/x\}$
(id)	$\vec{x}[M/\vec{x}]$	$=$	M
(ass$_1$)	$N[(M[L/\vec{y}])/\vec{x}]$	$=$	$(N[M/\vec{x}])[L/\vec{y}]$
(ass$_2$)	$(N[M/\vec{x}])[L/\vec{y}]$	$=$	$N[L \otimes M/\vec{x},\vec{y}]$
$(\otimes_1)$	$L \otimes (N[M/\vec{x}])$	$=$	$(L \otimes N)[M/\vec{x}]$
$(\otimes_2)$	$(M[L/\vec{x}]) \otimes N$	$=$	$(M \otimes N)[L/\vec{x}]$
(subst)	$(F(N))[M/\vec{x}]$	$=$	$F(N[M/\vec{x}])$

For instance, (ass$_1$) represents the associativity of substitutions. Other axioms are read similarly as specifying the expected properties of (suspended) substitutions.   □

Interestingly, α-conversion of the let-binding is derivable from our axioms.

Lemma 2.2.6 (substitution of multiple variables and α-conversion)

1. If $\Gamma,\vec{x}:\vec{\sigma} \vdash M : (\vec{\tau})$ and $\vec{y}$ do not occur in Γ,

$$\Gamma,\vec{y}:\vec{\sigma}_m \vdash \text{let } (\vec{x}) \text{ be } \vec{y} \text{ in } M = M\{\vec{y}/\vec{x}\} : (\vec{\tau}).$$

2. If $\Gamma \vdash M : (\vec{\sigma})$, $\Gamma,\vec{x}:\vec{\sigma} \vdash N : (\vec{\tau})$ and $\vec{y}$ do not occur in Γ nor in $\vec{x}$,

$$(\alpha) \quad \Gamma \vdash \text{let } (\vec{x}) \text{ be } M \text{ in } N = \text{let } (\vec{y}) \text{ be } M \text{ in } N\{\vec{y}/\vec{x}\} : (\vec{\tau}).$$

<u>Proof:</u>

1. The case of the number of variables m is 0:

$$
\begin{aligned}
& \text{let } () \text{ be } 0 \text{ in } M \\
= \quad & \text{let } () \text{ be } 0 \text{ in let } (\vec{x}) \text{ be } M \text{ in } \vec{x} \quad &\text{(id)} \\
= \quad & \text{let } (\vec{x}) \text{ be } M \text{ in } \vec{x} \quad &\text{(ass}_2) \\
= \quad & M \quad &\text{(id)}
\end{aligned}
$$

For $m \geq 1$, by induction on the number of variables:

$$
\begin{aligned}
& \text{let } (x_1,\ldots,x_{m+1}) \text{ be } y_1 \otimes \ldots y_{m+1} \text{ in } M \\
= \quad & \text{let } (x_1) \text{ be } y_1 \text{ in let } (x_2,\ldots,x_{m+1}) \text{ be } y_2 \otimes \ldots y_{m+1} \text{ in } M \quad &\text{(ass}_2) \\
= \quad & \text{let } (x_1) \text{ be } y_1 \text{ in } M\{y_2/x_2,\ldots,y_{m+1}/x_{m+1}\} \quad &\text{(hyp.)} \\
= \quad & M\{y_1/x_1,\ldots,y_{m+1}/x_{m+1}\} \quad &(\sigma_{\text{var}}).
\end{aligned}
$$

2.

$$
\begin{aligned}
& \text{let } (\vec{x}) \text{ be } M \text{ in } N \\
= \quad & \text{let } (\vec{x}) \text{ be } (\text{let } (\vec{y}) \text{ be } M \text{ in } \vec{y}) \text{ in } N \quad &\text{(id)} \\
= \quad & \text{let } (\vec{y}) \text{ be } M \text{ in let } (\vec{x}) \text{ be } \vec{y} \text{ in } N \quad &\text{(ass}_1) \\
= \quad & \text{let } (\vec{y}) \text{ be } M \text{ in } N\{\vec{y}/\vec{x}\} \quad &\text{(1.).} \quad \square
\end{aligned}
$$

Permutations of "independent" let-bindings are also justified:

Lemma 2.2.7 (permutation of let-bindings)
For $\Gamma \vdash L : (\vec{\sigma}_1)$, $\Gamma \vdash M : (\vec{\sigma}_2)$ and $\Gamma,\vec{x}:\vec{\sigma}_1,\vec{y}:\vec{\sigma}_2 \vdash N : (\vec{\tau})$,

$$\Gamma \vdash \text{let } (\vec{x}) \text{ be } L \text{ in let } (\vec{y}) \text{ be } M \text{ in } N = \text{let } (\vec{y}) \text{ be } M \text{ in let } (\vec{x}) \text{ be } L \text{ in } N : (\vec{\tau})$$

is derivable; we shall call this equation (perm).

<u>Proof:</u>

$$
\begin{aligned}
& \text{let } (\vec{x}) \text{ be } L \text{ in let } (\vec{y}) \text{ be } M \text{ in } N \\
= \quad & \text{let } (\vec{x}) \text{ be } L \text{ in let } (\vec{y}) \text{ be } M \text{ in let } (\vec{u},\vec{v}) \text{ be } \vec{x}\otimes\vec{y} \text{ in } N\{\vec{u}/\vec{x},\vec{v}/\vec{y}\} \quad &(\sigma_{\text{var}}) \\
= \quad & \text{let } (\vec{x}) \text{ be } L \text{ in let } (\vec{u},\vec{v}) \text{ be } (\text{let } (\vec{y}) \text{ be } M \text{ in } \vec{x}\otimes\vec{y}) \text{ in } N\{\vec{u}/\vec{x},\vec{v}/\vec{y}\} \quad &\text{(ass}_1) \\
= \quad & \text{let } (\vec{u},\vec{v}) \text{ be } (\text{let } (\vec{x}) \text{ be } L \text{ in let } (\vec{y}) \text{ be } M \text{ in } \vec{x}\otimes\vec{y}) \text{ in } N\{\vec{u}/\vec{x},\vec{v}/\vec{y}\} \quad &\text{(ass}_1) \\
= \quad & \text{let } (\vec{u},\vec{v}) \text{ be } (\text{let } (\vec{x}) \text{ be } L \text{ in } \vec{x}\otimes(\text{let } (\vec{y}) \text{ be } M \text{ in } \vec{y})) \text{ in } N\{\vec{u}/\vec{x},\vec{v}/\vec{y}\} \quad &(\otimes_2) \\
= \quad & \text{let } (\vec{u},\vec{v}) \text{ be } ((\text{let } (\vec{x}) \text{ be } L \text{ in } \vec{x})\otimes(\text{let } (\vec{y}) \text{ be } M \text{ in } \vec{y})) \text{ in } N\{\vec{u}/\vec{x},\vec{v}/\vec{y}\} \quad &(\otimes_1) \\
= \quad & \text{let } (\vec{u},\vec{v}) \text{ be } (\text{let } (\vec{y}) \text{ be } M \text{ in } (\text{let } (\vec{x}) \text{ be } L \text{ in } \vec{x})\otimes\vec{y}) \text{ in } N\{\vec{u}/\vec{x},\vec{v}/\vec{y}\} \quad &(\otimes_2) \\
= \quad & \text{let } (\vec{u},\vec{v}) \text{ be } (\text{let } (\vec{y}) \text{ be } M \text{ in let } (\vec{x}) \text{ be } L \text{ in } \vec{x}\otimes\vec{y}) \text{ in } N\{\vec{u}/\vec{x},\vec{v}/\vec{y}\} \quad &(\otimes_1) \\
= \quad & \text{let } (\vec{y}) \text{ be } M \text{ in let } (\vec{u},\vec{v}) \text{ be } (\text{let } (\vec{x}) \text{ be } L \text{ in } \vec{x}\otimes\vec{y}) \text{ in } N\{\vec{u}/\vec{x},\vec{v}/\vec{y}\} \quad &\text{(ass}_1) \\
= \quad & \text{let } (\vec{y}) \text{ be } M \text{ in let } (\vec{x}) \text{ be } L \text{ in let } (\vec{u},\vec{v}) \text{ be } \vec{x}\otimes\vec{y} \text{ in } N\{\vec{u}/\vec{x},\vec{v}/\vec{y}\} \quad &\text{(ass}_1) \\
= \quad & \text{let } (\vec{y}) \text{ be } M \text{ in let } (\vec{x}) \text{ be } L \text{ in } N \quad &(\sigma_{\text{var}})
\end{aligned}
$$

$\square$

To help with the intuition, we shall look at a few examples.

Example 2.2.8 As Example 2.1.5, we consider a sort $S = \{\texttt{nat}\}$ and a signature $\Sigma = \{\texttt{zero} : () \rightarrow (\texttt{nat}), \texttt{plus} : (\texttt{nat}, \texttt{nat}) \rightarrow (\texttt{nat})\}$. In an acyclic sharing theory over Σ, we can present the first sharing graph in Example 2.1.5 as

$$\vdash \textsf{let } (x) \textsf{ be zero in } \texttt{plus}(x \otimes x) : (\texttt{nat}).$$

It is easy to see that, in the pure acyclic sharing theory, this term is not equivalent to the "unshared" version

$$\vdash \texttt{plus}(\texttt{zero} \otimes \texttt{zero}) : (\texttt{nat})$$

because each axiom preserves the number of occurrences of operator symbols. However, as acyclic sharing theories lack the ability of representing cyclic bindings, we cannot present the second cyclic graph.   □

Example 2.2.9 In the example above, the terms can be expressed in many different ways. For instance,

$$
\begin{array}{lll}
& \texttt{plus}(\texttt{zero} \otimes \texttt{zero}) & \\
= & \textsf{let } (x, y) \textsf{ be zero} \otimes \texttt{zero in } \texttt{plus}(x \otimes y) & (\text{subst}) \\
= & \textsf{let } (x) \textsf{ be zero in let } (y) \textsf{ be zero in } \texttt{plus}(x \otimes y) & (\text{ass}_2) \\
= & \textsf{let } (x) \textsf{ be zero in } \texttt{plus}(\textsf{let } (y) \textsf{ be zero in } x \otimes y) & (\text{subst}) \\
= & \textsf{let } (x) \textsf{ be zero in } \texttt{plus}(x \otimes (\textsf{let } (y) \textsf{ be zero in } x \otimes y)) & (\otimes_1) \\
= & \textsf{let } (x) \textsf{ be zero in } \texttt{plus}(x \otimes \texttt{zero}) & (\text{id})
\end{array}
$$

and so on; we note that the third line is a normal form of this term (see below).   □

Example 2.2.10 One may wish to ignore isolated resources, for instance wanting to equate $\textsf{let } (x) \textsf{ be zero in } M$ with M if x is not free in M (thus the resource $\texttt{zero}$ is not referred from anywhere). This is not derivable in the pure acyclic sharing theory, and we need to assume an additional axiom (*garbage collection*):

$$M = 0 \textsf{ for any } \Gamma \vdash M : ()$$

Then, for instance, one can derive

$$
\begin{array}{lll}
& \textsf{let } (x) \textsf{ be zero in } M & \\
= & \textsf{let } () \textsf{ be } (\textsf{let } (x) \textsf{ be zero in } 0) \textsf{ in } M & \text{ass}_1 \\
= & \textsf{let } () \textsf{ be } 0 \textsf{ in } M & \text{garbage collection} \\
= & M.
\end{array}
$$

(Semantically, this condition amounts to assuming that $()$ is a terminal object. It is also called the affineness condition in the literature, c.f. [45, 60].)   □

Remark 2.2.11 Note that we have "empty bindings" for terms of type $()$, for example in $\textsf{let } () \textsf{ be } M \textsf{ in } N$ where M is of type $()$, as the case of the list of binding variables is empty. Actually there is no real binding of the subterm M in this expression; M can move around everywhere in the whole term.

$$
\begin{array}{lll}
\textsf{let } () \textsf{ be } M \textsf{ in } N = & \textsf{let } () \textsf{ be } M \textsf{ in let } (\vec{x}) \textsf{ be } N \textsf{ in } \vec{x} & (\text{id}) \\
= & \textsf{let } (\vec{x}) \textsf{ be } M \otimes N \textsf{ in } \vec{x} & (\text{ass}_2) \\
= & M \otimes N & (\text{id})
\end{array}
$$

Using (perm), one can prove that let () be M in $N = M \otimes N = N \otimes M$ for any $M : ()$. In particular, (the equivalence classes of) terms of type () form a commutative monoid.

□

Definition 2.2.12 (normal forms)
Normal forms are the well-typed terms generated from the following grammar.

$$n \ ::= \ \vec{x} \ | \ \text{let } (\vec{x}) \text{ be } F(\vec{y}) \text{ in } n$$

□

Theorem 2.2.13 (normal form theorem)
In the pure sharing theory, for any term $\Gamma \vdash M : (\vec{\tau})$, there is a normal form $\Gamma \vdash n : (\vec{\tau})$ such that $M = n$. Such a normal form is unique up to the congruence generated by (α) and (perm).

Proof: For existence, one may start from proving that, if n, n' are normal forms, then let $(\vec{x})$ be n in n' is equal to a normal form by the induction on the construction of n, taking care that substitution of variables preserves normal forms. Uniqueness follows from the fact that any axiom preserves operator symbols.   □

Remark 2.2.14 Our non-standard syntax arises from the existence of multiple conclusions (outputs). If we are interested just in single conclusions, then the following more standard syntax suffices.

$$M \ ::= \ x \ | \ F(M_1,\ldots,M_m) \ | \ \text{let } x \text{ be } M \text{ in } N$$

$$\frac{}{\Gamma, x : \sigma \vdash x : \sigma} \text{ variable}$$

$$\frac{\Gamma \vdash M_i : \sigma_i \quad (1 \le i \le m) \quad F : (\sigma_1,\ldots,\sigma_m) \to \tau}{\Gamma \vdash F(M_1,\ldots,M_m) : \tau} \text{ operator}$$

$$\frac{\Gamma \vdash M : \sigma \quad \Gamma, x : \sigma \vdash N : \tau}{\Gamma \vdash \text{let } x \text{ be } M \text{ in } N : \tau} \text{ let}$$

let x be y in M	$=$	$M\{y/x\}$
let x be M in x	$=$	M
let x be (let y be L in M) in N	$=$	let y be L in let x be M in N
let x_1 be M_1 in let x_2 be M_2 in N	$=$	let x_2 be M_2 in let x_1 be M_1 in N
let x be M in $F(\ldots,N,\ldots)$	$=$	$F(\ldots,\text{let } x \text{ be } M \text{ in } N,\ldots)$

It is routine to check that all axioms above are derivable from those of the pure acyclic sharing theory, thus there is a sound interpretation of this restricted version into the bigger theory. Moreover it is also easy to define the normal forms for this version, which are properly included in those for the pure acyclic sharing theory. So we conclude that the pure acyclic sharing theory is conservative over this restricted theory.

□

Relating Normal Forms and Sharing Graphs

We relate the pure acyclic sharing theory with finite acyclic sharing graphs as follows.

1. First, we show that acyclic graphs and normal forms are in bijective correspondence, up to renaming of bound variables (α-conversion) and permutations of operators which do not interfere with the dependency relation.

2. Since any term has a unique normal form up to α-conversion and (perm), we conclude that the acyclic sharing theory precisely axiomatizes finite acyclic sharing graphs.

Let $G = (V, L, A, (c_1, \ldots, c_n)) : (\sigma_1, \ldots, \sigma_m) \to (\tau_1, \ldots, \tau_n)$ be an acyclic sharing graph. Then it is possible to sort V's elements as $v_1, \ldots, v_k$ so that $i \leq j$ implies $v_j \not<^* v_i$. Of course this choice may not be unique, but we shall choose one of them, and build an expression of the pure acyclic sharing theory as

$$M_G \equiv \text{let } (x_{\langle v_1, 1\rangle}, \ldots, x_{\langle v_1, |\text{cod}(L(v_1))|\rangle}) \text{ be } L(v_1)(x_{A(v_1)_1}, \ldots, x_{A(v_1)_{|\text{dom}(L(v_1))|}}) \text{ in}$$

$$\ldots$$

$$\text{let } (x_{\langle v_k, 1\rangle}, \ldots, x_{\langle v_k, |\text{cod}(L(v_k))|\rangle}) \text{ be } L(v_k)(x_{A(v_k)_1}, \ldots, x_{A(v_k)_{|\text{dom}(L(v_k))|}}) \text{ in}$$

$$x_{c_1} \otimes \ldots \otimes x_{c_n}$$

where we introduce a variable x_c for each $c \in C$ (recall the definition). We claim that M_G is a well-typed expression (hence a normal form):

Lemma 2.2.15 $x_{d_1} : \sigma_1, \ldots, x_{d_m} : \sigma_m \vdash M_G : (\tau_1, \ldots, \tau_n)$.

<u>Proof:</u> Induction on the size of V. If V is empty, M_G is simply $x_{c_1} \otimes \ldots \otimes x_{c_n}$, and since each c_i is contained in $\{d_1, \ldots, d_m\}$ M_G's typing is derivable from the rules (variable) and (tensor) (if $n > 0$) or (unit) (if $n = 0$). If V is not empty, one may consider a sharing graph $G' = (V', A', L', (c_1, \ldots, c_n)) : (\sigma_1, \ldots, \sigma_{m+|\text{dom}(L(v_1))|}) \to (\tau_1, \ldots, \tau_n)$ where $V' = V - \{v_1\}$, $L' = L|_{V'}$, $A'(v)_i = d_{m+j}$ if $A'(v)_i = \langle v_1, j\rangle$ and $A'(v)_i = A(v)_i$ otherwise. σ_{m+i} is $\text{dom}(L(v_1))_i$. Then (by the same linear-ordering)

$$M_G \equiv \text{let } (x_{\langle v_1, 1\rangle}, \ldots, x_{\langle v_1, |\text{cod}(L(v_1))|\rangle})$$
$$\text{be } L(v_1)(x_{A(v_1)_1}, \ldots, x_{A(v_1)_{|\text{dom}(L(v_1))|}})$$
$$\text{in } M_{G'}\{x_{\langle v_1, j\rangle}/x_{d_{m+j}}\}$$

On the other hand, by the induction hypothesis,

$$x_{d_1} : \sigma_1, \ldots, x_{d_{m+|\text{dom}(L(v_1))|}} : \sigma_{m+|\text{dom}(L(v_1))|} \vdash M'_G : (\tau_1, \ldots, \tau_n)$$

By the rule (operator), (let) and renaming of variables by $x_{d_{m+j}} \mapsto x_{\langle v_1, j\rangle}$ we obtain the typing of M_G.   □

By assuming another linear ordering of V's elements, we get another well-typed expression M'_G. However, the difference between M_G and M'_G lies only in the permutation of the let-bindings, which is guaranteed by (perm) (routinely shown by induction on the number of nodes). Thus $M_G = M'_G$ is provable in the sharing calculus. Also it is routine to check that equivalent sharing graphs give equivalent normal forms. Conversely, a normal form n gives rise to a sharing graph G_n, by reversing the construction above, such that $G_{M_G} \equiv G$ and $M_{G_n} = n$. Together with the proposition of normal forms, we have

Theorem 2.2.16 There is a bijective correspondence between the finite acyclic sharing graphs and the equivalence classes of terms of the pure acyclic sharing theory. □

Actually this should be more than a bijective correspondence; this must preserve the algebraic nature behind sharing graphs, though at this moment it is not very clear. The leading paradigm in this thesis is that "sharing graphs (and their models) form a nice categorical structure" – this bijection turns out to respect such semantic structure behind sharing graphs (see Remark 3.3.8). While not many work in the literature spot on the structural nature of sharing graphs, we shall point out that Milner's treatment of his "molecular forms" as an action structure (symmetric monoidal category with structure) [68] certainly lies in this direction, and we acknowledge the influence of his work on ours.

2.3 Cyclic Sharing Theory

In a parallel manner to the acyclic sharing graphs and acyclic sharing theory, we give the equational theory for cyclic sharing graphs, called *cyclic sharing theory*. Ignoring small technical points, all we need is to replace the let-bindings (acyclic bindings) by the letrec-bindings (cyclic bindings). The syntax is close to that of many applicative programming languages with recursive bindings, although at this moment our theory corresponds to the cyclic sharing graphs and says nothing about (lazy) recursive computation as often represented by the letrec-syntax in lazy functional languages.

Definition 2.3.1 (raw expressions)

$$M ::= x \mid F(M) \mid 0 \mid M_1 \otimes M_2 \mid \text{letrec } (x_1, \ldots, x_m) \text{ be } M_1 \text{ in } M_2$$

We identify $M \otimes 0$ and $0 \otimes M$ with M, and $(M_1 \otimes M_2) \otimes M_3$ with $M_1 \otimes (M_2 \otimes M_3)$.

Definition 2.3.2 (typing)

$$\frac{}{\Gamma, x : \sigma \vdash x : (\sigma)} \text{ variable}$$

$$\frac{\Gamma \vdash M : (\sigma_1, \ldots, \sigma_m) \quad F : (\sigma_1, \ldots, \sigma_m) \to (\tau_1, \ldots, \tau_n)}{\Gamma \vdash F(M) : (\tau_1, \ldots, \tau_n)} \text{ operator}$$

$$\frac{}{\Gamma \vdash 0 : ()} \text{ unit}$$

$$\frac{\Gamma \vdash M : (\sigma_1, \ldots, \sigma_m) \quad \Gamma \vdash N : (\tau_1, \ldots, \tau_n)}{\Gamma \vdash M \otimes N : (\sigma_1, \ldots, \sigma_m, \tau_1, \ldots, \tau_n)} \text{ tensor}$$

$$\frac{\Gamma, x_1 : \sigma_1, \ldots, x_m : \sigma_m \vdash M : (\sigma_1, \ldots, \sigma_m) \qquad \Gamma, x_1 : \sigma_1, \ldots, x_m : \sigma_m \vdash N : (\tau_1, \ldots, \tau_n)}{\Gamma \vdash \text{letrec } (x_1, \ldots, x_m) \text{ be } M \text{ in } N : (\tau_1, \ldots, \tau_n)} \text{ letrec}$$

$$\frac{\Gamma, x : \sigma, x' : \sigma', \Gamma' \vdash M : (\tau_1, \ldots, \tau_n)}{\Gamma, x' : \sigma', x : \sigma, \Gamma' \vdash M : (\tau_1, \ldots, \tau_n)} \text{ exchange}$$

□

Definition 2.3.3 (axioms)

(σ_{var})	letrec $(x, \vec{y})$ be $z \otimes M$ in N	$=$	letrec $(\vec{y})$ be $M\{z/x\}$ in $N\{z/x\}$ $(x \not\equiv z)$
(id)	letrec $(\vec{x})$ be M in $\vec{x}$	$=$	M
(ass_1)	letrec $(\vec{x})$ be (letrec $(\vec{y})$ be L in M) in N	$=$	letrec $(\vec{x}, \vec{y})$ be $M \otimes L$ in N
(ass_2)	letrec $(\vec{x})$ be L in letrec $(\vec{y})$ be M in N	$=$	letrec $(\vec{x}, \vec{y})$ be $L \otimes M$ in N
$(\otimes_1)$	$L \otimes$ (letrec $(\vec{x})$ be M in N)	$=$	letrec $(\vec{x})$ be M in $L \otimes N$
$(\otimes_2)$	(letrec $(\vec{x})$ be L in M) $\otimes N$	$=$	letrec $(\vec{x})$ be L in $M \otimes N$
(perm)	letrec $(\vec{x}, \vec{y}, \vec{z})$ be $M_1 \otimes M_2 \otimes M_3$ in N	$=$	letrec $(\vec{y}, \vec{x}, \vec{z})$ be $M_2 \otimes M_1 \otimes M_3$ in N
(subst)	letrec $(\vec{x})$ be M in $F(N)$	$=$	$F(\text{letrec } (\vec{x}) \text{ be } M \text{ in } N)$

As the acyclic case, both sides of axioms must have the same type under the same context. For instance, in (id), $\vec{x}$ cannot be free in M. □

Definition 2.3.4 (cyclic sharing theory)
A *cyclic sharing theory* over Σ is an equational theory on the well-typed terms closed under the term construction described above, where the equality on terms is a congruence relation containing the axioms above; the only difference from that of an acyclic sharing theory is the following inference rule for letrec is used instead of that for let.

$$\frac{\Gamma, \vec{x} : \vec{\sigma} \vdash M = M' : (\vec{\sigma}) \qquad \Gamma, \vec{x} : \vec{\sigma} \vdash N = N' : (\vec{\tau})}{\Gamma \vdash \text{letrec } (\vec{x}) \text{ be } M \text{ in } N = \text{letrec } (\vec{x}) \text{ be } M' \text{ in } N' : (\vec{\tau})}$$

By the *pure cyclic sharing theory*, we mean the cyclic sharing theory with no additional axioms. □

Again we give a pictorial account for the letrec-construction:

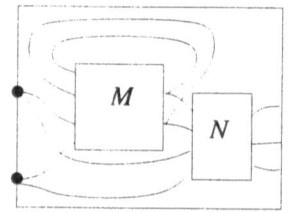

$$\frac{\Gamma, \vec{x} : \vec{\sigma} \vdash M : (\vec{\sigma}) \qquad \Gamma, \vec{x} : \vec{\sigma} \vdash N : (\vec{\tau})}{\Gamma \vdash \text{letrec } (\vec{x}) \text{ be } M \text{ in } N : (\vec{\tau})}$$

For readability, we introduce the following syntax for the multiple letrec-binding:

$$\frac{\Gamma,\vec{x}_1 : \vec{\sigma}_1,\ldots,\vec{x}_k : \vec{\sigma}_k \vdash M_i : (\vec{\sigma}_i) \quad (1 \leq i \leq k)}{\Gamma,\vec{x}_1 : \vec{\sigma}_1,\ldots,\vec{x}_k : \vec{\sigma}_k \vdash N : (\vec{\tau})}$$
$$\overline{\Gamma \vdash \text{letrec } (\vec{x}_1) \text{ be } M_1,\ldots (\vec{x}_k) \text{ be } M_k \text{ in } N : (\vec{\tau})}$$

for letrec $(\vec{x}_1,\ldots,\vec{x}_k)$ be $M_1 \otimes \ldots \otimes M_k$ in N. For instance, the axiom (perm) is equivalent to

letrec ... $(\vec{x})$ be M, $(\vec{y})$ be M' ... in N = letrec ... $(\vec{y})$ be M', $(\vec{x})$ be M ... in N

which will be referred as (perm'). Also, the axiom (σ_{var}) is the same as

letrec (x) be z, $(\vec{y})$ be M in N = letrec $(\vec{y})$ be $M\{z/x\}$ in $N\{z/x\}$

where $x \not\equiv z$. It is easy but helpful to see that this can be replaced by a pair of axioms representing "dereference" and "garbage collection" (restricted on variables):

(deref) letrec (x) be z, $(\vec{y})$ be M in N = letrec (x) be z, $(\vec{y})$ be $M\{z/x\}$ in $N\{z/x\}$
(g.c.) letrec (x) be z in N = N $(x \notin FV(N)$ & $x \not\equiv z)$

Note that the garbage collection of trivial cycles is not allowed (in the pure cyclic sharing theory), e.g. letrec (x) be x in M = M is not derivable even when x is not free in M. This alternative axiomatization will be used when we consider the extension with higher-order constructs where substitutions of values are allowed.

As in the acyclic theory, α-conversion of letrec-bound variables is derivable.

Lemma 2.3.5 (α-conversion)
If $\Gamma \vdash M : (\vec{\sigma})$, $\Gamma,\vec{x} : \vec{\sigma} \vdash N : (\vec{\tau})$ and $\vec{y}$ do not occur in Γ nor in $\vec{x}$,

(α) $\Gamma \vdash$ letrec $(\vec{x})$ be M in N = letrec $(\vec{y})$ be $M\{\vec{y}/\vec{x}\}$ in $N\{\vec{y}/\vec{x}\} : (\vec{\tau})$.

Proof:

       letrec $(\vec{x})$ be M in N
=   letrec $(\vec{x})$ be (letrec $(\vec{y})$ be M in $\vec{y}$) in N (id)
=   letrec $(\vec{x},\vec{y})$ be $\vec{y} \otimes M$ in N (ass$_1$)
=   letrec $(\vec{y})$ be $M\{\vec{y}/\vec{x}\}$ in $N\{\vec{y}/\vec{x}\}$ (σ_{var}).

$\square$

The difference between the axiomatizations of acyclic sharing theory and the cyclic one is summarized as follows.

- The axiom (σ_{var}) is changed to so that a substitution is effective throughout under the scope of the letrec-binding. The acyclic version is too weak, for instance for deriving letrec (x,y) be $y \otimes x$ in y = letrec (y) be y in y. However, there is no other difference; z must be different from x, so we cannot use this axiom for deleting a trivial cycle as mentioned above.

- The axiom (perm) is derivable in the acyclic theory, but it is not the case for the cyclic theory, because of the cyclic bindings. In (perm), M_i's can depend on each other, i.e. can contain any free x, y and z's, so the proof of (perm) for the acyclic theory using (ass_1) and $(\otimes_i)$ cannot be applied.

- The axiom (ass_1) is slightly more general than that for the acyclic theory for free xs can occur in L and M.

However, the other axioms (id), (ass_2), $(\otimes_i)$ and (subst) are unchanged.

Example 2.3.6 The second graph of Example 2.1.5 can be presented as

$$\vdash \text{letrec } (x) \text{ be plus}(\text{zero}, x) \text{ in } x : (\text{nat}).$$

$\square$

Example 2.3.7 The tangled graph of Example 2.1.7 can be presented as

$$x : \text{bool}, y : \text{nat} \vdash \quad \text{letrec} \quad (z, z') \text{ be } F(z', y),$$
$$(u, u') \text{ be } F(x, v),$$
$$(v, v') \text{ be } F(u', w),$$
$$(w) \text{ be } w$$
$$\text{in} \quad u \otimes v \otimes v' \qquad : \text{nat} \otimes \text{nat} \otimes \text{bool}.$$

$\square$

For defining the notion of normal forms, we need a care about trivial cycles again.

Definition 2.3.8 For each sort σ, define $\bullet^\sigma : (\sigma)$ by $\bullet^\sigma \equiv \text{letrec } (x) \text{ be } x \text{ in } x.$ $\square$

Definition 2.3.9 (normal forms)
Normal forms are the well-typed terms of the following form

$$\text{letrec } (\vec{x}_1) \text{ be } F_1(\vec{y}_1) \text{ in } \ldots (\vec{x}_m) \text{ be } F_m(\vec{y}_m) \text{ in } \vec{z}$$

where F_i is either an operator symbol or $\bullet$. $\square$

Theorem 2.3.10 (normal form theorem)
In the pure cyclic sharing theory, for any term $\Gamma \vdash M : (\vec{t})$, there is a normal form $\Gamma \vdash n : (\vec{t})$ such that $M = n$. Such a normal form is unique up to the congruence generated by (α) and (perm').

Proof: Similar to the case of the pure acyclic sharing theory. (N.B. [67] contains the essentially same result on reflexive action calculi and molecular forms.) $\square$

It is easy to give a cyclic sharing graph from a normal form and vise versa – unlike the acyclic case, we do not have to care about the dependency between nodes, and it is routine to show

Theorem 2.3.11 There is a bijective correspondence between the finite cyclic sharing graphs and the equivalence classes of terms of the pure cyclic sharing theory. □

Again this should be more than a bijective correspondence; this must preserve the algebraic nature implicit in cyclic sharing graphs. This becomes the central issue in the later development of the semantic counterpart.

Remark 2.3.12 The same remark applies as in the acyclic case: if we are interested just in the single conclusion, then the following more standard syntax suffices.

Raw Terms	M	$::=$ $x \mid F(M_1,\ldots,M_m) \mid$ letrec x_1 be $M_1,\ldots,x_m$ be M_m in N
Declarations	D	$::=$ x be $M \mid x$ be M,D

In a declaration, binding variables are assumed to be disjoint.

$$\frac{}{\Gamma,x:\sigma \vdash x:\sigma} \text{ variable}$$

$$\frac{\Gamma \vdash M_i : \sigma_i \quad (1 \leq i \leq m) \quad F : (\sigma_1,\ldots,\sigma_m) \to \tau}{\Gamma \vdash F(M_1,\ldots,M_m) : \tau} \text{ operator}$$

$$\frac{\Gamma,x_1:\sigma_1,\ldots,x_m:\sigma_m \vdash M_i:\sigma_i \quad (1 \leq i \leq m) \quad \Gamma,x_1:\sigma_1,\ldots,x_m:\sigma_m \vdash N:\tau}{\Gamma \vdash \text{letrec } x_1 \text{ be } M_1,\ldots,x_m \text{ be } M_m \text{ in } N:\tau} \text{ letrec}$$

letrec x be z,D in M	$=$	letrec $D\{z/x\}$ in $M\{z/x\}$ $(x \not\equiv z)$
letrec x be M in x	$=$	M
letrec y be (letrec D_1 in M),D_2 in N	$=$	letrec D_1,y be M,D_2 in N
letrec D_1 in letrec D_2 in M	$=$	letrec D_1,D_2 in M
letrec D_1,D_2,D_3 in N	$=$	letrec D_2,D_1,D_3 in N
letrec x be M in $F(\ldots,N,\ldots)$	$=$	$F(\ldots,\text{letrec } x \text{ be } M \text{ in } N,\ldots)$

□

2.4 Rewriting on Sharing Graphs

Rewriting is the computational counterpart of sharing graphs. There is a wide variety of formulations of rewriting on sharing graphs from abstract ones to concrete implementation oriented ones [84], but here we choose an intermediate approach, which is close to Ariola and Klop's *equational term graph rewriting*.

Definition 2.4.1 (rewriting system)
Let $\mathbb{T}$ be a (acyclic or cyclic) sharing theory. A *rewriting system on* $\mathbb{T}$ is a preorder on the equivalence classes of well-typed terms with the same contexts and same types, closed under all term constructions. □

Spelling this out, a rewriting system can be described as a family of relations $\succ_{\Gamma,(\vec{\sigma})}$ for each context Γ and type $(\vec{\sigma})$ satisfying the following conditions.

- $\succ_{\Gamma,(\vec{\sigma})}$ is a binary relation on well-typed terms with context Γ and type $(\vec{\sigma})$ (subscripts may be omitted). We write $\Gamma \vdash M \succ N : (\vec{\sigma})$ for $M \succ_{\Gamma,(\vec{\sigma})} N$.

- $\succ_{\Gamma,(\vec{\sigma})}$ respects equality:

$$\frac{\Gamma \vdash M = N : (\vec{\sigma})}{\Gamma \vdash M \succ N : (\vec{\sigma})}$$

- $\succ_{\Gamma,(\vec{\sigma})}$ is a preorder:

$$\frac{\Gamma \vdash M : (\vec{\sigma})}{\Gamma \vdash M \succ M : (\vec{\sigma})} \qquad \frac{\Gamma \vdash L \succ M : (\vec{\sigma}) \quad \Gamma \vdash M \succ N : (\vec{\sigma})}{\Gamma \vdash L \succ N : (\vec{\sigma})}$$

- Closure under term constructions:

$$\frac{\Gamma \vdash M \succ N : (\vec{\sigma}) \quad F : (\vec{\sigma}) \to (\vec{\tau})}{\Gamma \vdash F(M) \succ F(N) : (\vec{\tau})} \qquad \frac{\Gamma \vdash M \succ M' : (\vec{\sigma}) \quad \Gamma \vdash N \succ N' : (\vec{\tau})}{\Gamma \vdash M \otimes N \succ M' \otimes N' : (\vec{\sigma}, \vec{\tau})}$$

$$\frac{\Gamma, x : \sigma, x' : \sigma', \Gamma' \vdash M \succ N : (\vec{\tau})}{\Gamma, x' : \sigma', x : \sigma, \Gamma' \vdash M \succ N : (\vec{\tau})}$$

For acyclic theory

$$\frac{\Gamma \vdash M \succ M' : (\vec{\sigma}) \quad \Gamma, \vec{x} : \vec{\sigma} \vdash N \succ N' : (\vec{\tau})}{\Gamma \vdash \text{let } (\vec{x}) \text{ be } M \text{ in } N \succ \text{let } (\vec{x}) \text{ be } M' \text{ in } N' : (\vec{\tau})}$$

For cyclic theory

$$\frac{\Gamma, \vec{x} : \vec{\sigma} \vdash M \succ M' : (\vec{\sigma}) \quad \Gamma, \vec{x} : \vec{\sigma} \vdash N \succ N' : (\vec{\tau})}{\Gamma \vdash \text{letrec } (\vec{x}) \text{ be } M \text{ in } N \succ \text{letrec } (\vec{x}) \text{ be } M' \text{ in } N' : (\vec{\tau}).}$$

This notion of rewriting is too rough. In most practical settings, it is unlikely that some resource is created from nothing, but our definition allows such a stupid rewriting system! Thus we are led to restrict the notion of rewriting in an adequate way:

Definition 2.4.2 (minimality condition)
A rewriting system satisfies the *minimality condition* if $\Gamma \vdash \vec{x} \succ M : (\vec{\sigma})$ implies $\Gamma \vdash M = \vec{x} : (\vec{\sigma})$. □

Example 2.4.3 The famous *cyclic I problem* can be a good test case for a formalization of graph rewriting systems. Let us consider the pure cyclic sharing theory with an operator symbol $I : (\sigma) \to (\sigma)$ and a rewriting system generated by a rule $I(x) \succ x$, which satisfies the minimality condition. Now we shall look at a cyclic term $C \equiv \text{letrec } (x) \text{ be } I(x) \text{ in } x$. How can we rewrite this C? This is not a trivial question as it first looks like. An early attempt (Barendregt et al. [15]) was close to the traditional term rewriting. Following the infinite unwinding semantics, one may regard this as a representation of the infinite application of I, i.e. $C = I(C) = I(I(C)) = \ldots = I(I(I\ldots))$. Having this "unwinding" $C = I(C)$ in mind,

$$C = \underline{I(C)} \succ C$$

therefore rewrite C to itself. However it was pointed out that this solution causes non-confluency (consider letrec (x) be $F(G(x))$ in x with rules $F(x) \succ x$ and $G(x) \succ x$), see e.g. [11]. To get rid of this problem, a new solution was proposed in [11], respecting the circularly shared structure of C:

$$C \equiv \text{letrec } (x) \text{ be } \underline{I(x)} \text{ in } x \succ \text{letrec } (x) \text{ be } x \text{ in } x \equiv \bullet$$

Thus we rewrite C to the blackhole constant $\bullet$. Our rewriting system is this new version. Indeed our equational theory does not equate C to $I(C)$.

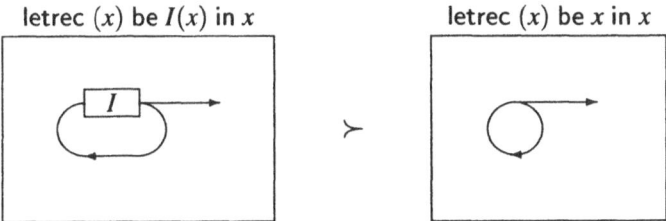

letrec (x) be $I(x)$ in x letrec (x) be x in x

$\square$

2.5 Equational Term Graph Rewriting

We conclude this chapter by indicating the connection between the *equational term graph rewriting* of Ariola and Klop [11] (see also Ariola and Arvind [7]) and our equational treatment of sharing graphs which was largely inspired by their work.

The equational theory underlying Ariola and Klop's work is, in spirit, essentially identical to that of ours – but not quite the same; see the discussion below. Our equational theory allows nestings of systems of equations (let/letrec-blocks) whereas theirs does not, but this is a matter of presentation. We have chosen our richer version to make the structural (algebraic) nature of sharing graphs clear, and this choice can be justified by the semantic considerations throughout the rest of this thesis.

The significance of the Ariola-Klop approach lies on the clear treatment of the bisimilarity between systems of equations and some basic operations on them, including copying, substitution and flattening. However, while they treat these operations as rewriting steps on systems of equations and have shown the confluence results, the connection with real rewriting on sharing graphs is not always simple, as some of these operations do not change the corresponding sharing graphs (and some do).

Our rewriting systems are designed not on systems of equations but on equivalence classes of them, so specify rewriting on sharing graphs themselves. In this sense our choice is more abstract; the copying and substitution operations are part of our equational theory as long as they do not change the corresponding graph; and the flattening is contained in our equational theory.

We believe that, while Ariola-Klop's results are elegant, proper computation on sharing graphs must be represented by the rewriting systems on them, and our "up to equivalence classes" (which does mean "up to sharing graphs") approach is suitable for this purpose. Also, for discussing the semantic contents of rewriting systems, it is important to separate the rewriting on representations (systems of equations) from that on the real semantic objects (sharing graphs) strictly. We hope that our direction will be more justified by further research.

3

Models of Acyclic Sharing Theory

This chapter presents the most basic part of our development of the semantic models of sharing graphs.

Here we deal with just the acyclic case. The reader may be disappointed, as technically and practically (and conceptually!) cyclic sharing graphs are far more interesting, and also the difference between acyclic and cyclic settings are relatively small if we just look at the definition of sharing graphs - in fact we define acyclic graphs as special instances of cyclic ones, so one may wonder why we do not start our semantic story from the models of cyclic sharing.

However, it turns out that the models of acyclic sharing theories admit a fairly simple category-theoretic formulation, which can be regarded as a generalization of the traditional treatment of algebraic theories in categorical type theory. And, as any other correspondence between syntactic theories and semantic models in mathematics and computer science, allowing stronger constructions (in this case cyclic sharing) amounts to restricting the class of models by assuming additional conditions. Starting from the models of acyclic sharing, which are fairly flexible and general, in later chapters we add new conditions to make them models of higher-order sharing graphs (the condition will be a form of *adjunction*), as well as models of cyclic sharing (the condition is a new notion called *trace*).

We believe that this systematic development is useful to identify what is needed for modeling (i.e. implementing!) sharing graphs, in a step-by-step manner. Also we think it useful to make the comparison with the traditional approach to algebraic theories as clear as we can. So, though computationally not very compelling, we describe our theory-model correspondence in the standard style of categorical type theory as found in [26].

The category theory needed in this thesis is not heavy at all, but we have to use the language for monoidal categories very frequently. So we shall first summarize the material which will be used throughout the rest of this thesis. After that, we introduce our class of models, and show the expected properties such as soundness, completeness and the theory-model correspondence. Also we observe that rewriting systems on sharing theories have the semantic interpretation in our models as local preorders.

3.1 Preliminaries from Category Theory

We shall briefly recall basic definitions of monoidal categories [59, 49] which will be the main technical language in the rest of this thesis.

Definition 3.1.1 (monoidal categories)

A *monoidal category* (or *tensor category*) $\mathcal{M} = (\mathcal{M}, \otimes, I, a, l, r)$ consists of a category $\mathcal{M}$, a functor $\otimes : \mathcal{M} \times \mathcal{M} \to \mathcal{M}$ (called the *monoidal* or *tensor product*), an object $I \in \mathcal{M}$ (the *unit object*) and natural isomorphisms

$$
\begin{array}{lll}
a_{A,B,C} & : & (A \otimes B) \otimes C \xrightarrow{\sim} A \otimes (B \otimes C) \\
l_A & : & I \otimes A \xrightarrow{\sim} A \\
r_A & : & A \otimes I \xrightarrow{\sim} A
\end{array}
$$

such that, for objects $A, B, C, D \in \mathcal{M}$, the following two diagrams (called the *associativity pentagon* and the *triangle for unit* respectively) commute:

$$
\begin{array}{ccc}
((A \otimes B) \otimes C) \otimes D & \xrightarrow{\quad\quad a \quad\quad} & (A \otimes B) \otimes (C \otimes D) \\
\downarrow{\scriptstyle a \otimes D} & & \downarrow{\scriptstyle a} \\
(A \otimes (B \otimes C)) \otimes D \xrightarrow{a} A \otimes ((B \otimes C) \otimes D) & \xrightarrow{A \otimes a} & A \otimes (B \otimes (C \otimes D))
\end{array}
$$

$$
\begin{array}{ccc}
(A \otimes I) \otimes B & \xrightarrow{\quad a \quad} & A \otimes (I \otimes B) \\
& {\scriptstyle r \otimes B} \searrow \quad \swarrow {\scriptstyle A \otimes l} & \\
& A \otimes B &
\end{array}
$$

A *strict monoidal category* is a monoidal category in which all of $a_{A,B,C}$, l_A and r_A are identity arrows (hence $A \otimes (B \otimes C)$ and $(A \otimes B) \otimes C$ are the same object, and so are $A \otimes I, I \otimes A$ and A).   □

Definition 3.1.2 (symmetric monoidal categories)

A *symmetry* for a monoidal category is a natural transformation

$$c_{A,B} : A \otimes B \to B \otimes A$$

subject to the following two commutative diagrams (the *bilinearity* and *symmetry*):

$$
\begin{array}{ccccc}
(A \otimes B) \otimes C & \xrightarrow{\ a\ } & A \otimes (B \otimes C) & \xrightarrow{\ c\ } & (B \otimes C) \otimes A \\
\downarrow{\scriptstyle c \otimes C} & & & & \downarrow{\scriptstyle a} \\
(B \otimes A) \otimes C & \xrightarrow{\ a\ } & B \otimes (A \otimes C) & \xrightarrow{B \otimes c} & B \otimes (C \otimes A)
\end{array}
$$

$$
\begin{array}{ccc}
A \otimes B & & \\
\downarrow{\scriptstyle c} & \searrow {\scriptstyle A \otimes B} & \\
B \otimes A & \xrightarrow{\ c\ } & A \otimes B
\end{array}
$$

Note that the symmetry condition implies that c is a natural isomorphism. A monoidal category equipped with a symmetry is called a *symmetric monoidal category (SMC)*. A *strict* symmetric monoidal category is then a strict monoidal category with symmetry – note that the symmetry need not be "strict", as we do not assume that $A \otimes B = B \otimes A$.

$\square$

Example 3.1.3 Here are some examples of monoidal categories:

- The category of sets and partial functions is symmetric monoidal, where the monoidal product is given by the direct product.

- Similarly, the category of sets and binary relations is symmetric monoidal with the direct product as the monoidal product.

- Abelian groups and homomorphisms form a symmetric monoidal category; the unit object is $\mathbb{Z}$ and the monoidal product is the tensor product of abelian groups.

- Let C be a category. The category $[C, C]$ has endofunctors on C as objects and natural transformations between them as arrows. The composition of functors gives a strict monoidal product on $[C, C]$, where the identity functor serves as the unit object. This monoidal category may not have a symmetry. $\square$

In this thesis, we frequently use strict symmetric monoidal categories naturally arising from syntactic constructions. For ease of later developments, we shall give an axiomatization of strict symmetric monoidal categories below (the labels of axioms are purely conventional).

[axioms for strict monoidal category]

M1	$f; id = f = id; f$	M4 $f; (g; h) = (f; g); h$
M2	$f \otimes id_I = f = id_I \otimes f$	M5 $f \otimes (g \otimes h) = (f \otimes g) \otimes h$
M3	$id \otimes id = id$	M6 $(f; g) \otimes (f'; g') = (f \otimes f'); (g \otimes g')$

[axioms for symmetry]

S1	$c; (f \otimes g) = (g \otimes f); c$	S3 $c_{A,B}; c_{B,A} = id_{A \otimes B}$
S2	$(c_{A,B} \otimes id_C); (id_B \otimes c_{A,C}) = c_{A, B \otimes C}$	

Example 3.1.4 (cartesian categories)
In this thesis, by a *cartesian category* we mean a category with *chosen* binary products (binary product cones) and a *chosen* terminal object. We write $A \times B$ for the chosen binary product of objects A and B, and also write $A \overset{\pi_{A,B}}{\leftarrow} A \times B \overset{\pi'_{A,B}}{\rightarrow} B$ for the chosen product cone. Given a cone $A \overset{f}{\leftarrow} C \overset{g}{\rightarrow} B$, $\langle f, g \rangle : C \to A \times B$ is then the uniquely determined arrow such that $\langle f, g \rangle; \pi_{A,B} = f$ and $\langle f, g \rangle; \pi'_{A,B} = g$. The chosen terminal object will be denoted by 1, and the terminal map is $!_A : A \to 1$.

From this information, a cartesian category is regarded as a symmetric monoidal category whose monoidal products are binary products and unit object is the terminal object:

$$\begin{aligned}
f \times g &= \langle \pi_{A,B}; f, \pi'_{A,B}; g \rangle \quad \text{for } f : A \to A' \text{ and } g : B \to B' \\
a_{A,B,C} &= \langle \pi_{A \times B,C}; \pi_{A,B}, \langle \pi_{A \times B,C}; \pi'_{A,B}, \pi'_{A \times B,C} \rangle \rangle \\
l_A &= \pi'_{1,A} \\
r_A &= \pi_{A,1} \\
c_{A,B} &= \langle \pi'_{A,B}, \pi_{A,B} \rangle
\end{aligned}$$

A *strict cartesian category* is a cartesian category in which all components of a, l and r are identity arrows. □

Definition 3.1.5 (monoidal functors [30])
Given monoidal categories $\mathcal{M} = (\mathcal{M}, \otimes, I, a, l, r)$ and $\mathcal{M}' = (\mathcal{M}', \otimes', I', a', l', r')$, a *monoidal functor* from $\mathcal{M}$ to $\mathcal{M}'$ is a tuple (F, m, m_I) where F is a functor from $\mathcal{M}$ to $\mathcal{M}'$, m is a natural transformation from $F(-) \otimes' F(=)$ to $F(- \otimes =)$ and $m_I : I' \to FI$ in $\mathcal{M}'$ which satisfy the coherence conditions below.

$$\begin{array}{ccccc}
(FA \otimes' FB) \otimes' FC & \xrightarrow{m \otimes' FC} & F(A \otimes B) \otimes' FC & \xrightarrow{m} & F((A \otimes B) \otimes C) \\
\Big\downarrow{a'} & & & & \Big\downarrow{Fa} \\
FA \otimes' (FB \otimes' FC) & \xrightarrow[FA \otimes' m]{} & FA \otimes' F(B \otimes C) & \xrightarrow[m]{} & F(A \otimes (B \otimes C))
\end{array}$$

$$\begin{array}{ccc}
I' \otimes' FA & \xrightarrow{l'} & FA \\
\Big\downarrow{m_I \otimes FA} & & \Big\uparrow{Fl} \\
FI \otimes' FA & \xrightarrow[m]{} & F(I \otimes A)
\end{array}
\qquad
\begin{array}{ccc}
FA \otimes' I' & \xrightarrow{r'} & FA \\
\Big\downarrow{FA \otimes m_I} & & \Big\uparrow{Fr} \\
FA \otimes' FI & \xrightarrow[m]{} & F(A \otimes I)
\end{array}$$

□

Definition 3.1.6 (symmetric monoidal functor [30])
Let $\mathcal{M} = (\mathcal{M}, \otimes, I, a, l, r, c)$ and $\mathcal{M}' = (\mathcal{M}', \otimes', I', a', l', r', c')$ be symmetric monoidal categories. A *symmetric monoidal functor* from $\mathcal{M}$ to $\mathcal{M}'$ is a monoidal functor (F, m, m_I) which additionally satisfies the following condition:

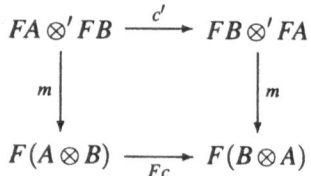

□

Lemma 3.1.7 (composition of monoidal functors [30])
For (symmetric) monoidal functors $(F, m, m_I) : \mathcal{M} \to \mathcal{M}'$ and $(G, n, n_{I'}) : \mathcal{M}' \to \mathcal{M}''$,

$(G, n, n_{I'}) \circ (F, m, m_I) \equiv (G \circ F, G(m) \circ n_{F,F}, G(m_I) \circ n_{I'})$ is a (symmetric) monoidal functor from $\mathcal{M}$ to $\mathcal{M}'$. This composition is associative, and satisfies the identity law for the identity (symmetric) monoidal functor. □

Definition 3.1.8 (strong/strict monoidal functors)

- A monoidal functor (F, m, m_I) is called *strong* if m is a natural isomorphism and m_I is an isomorphism.

- A monoidal functor (F, m, m_I) is called *strict* if all components of m and m_I are identity arrows. □

Example 3.1.9 (finite product preserving functors)
Let C be a cartesian category and $\mathcal{D}$ a category. We say that a functor $F : C \to \mathcal{D}$ *preserves finite products* if F sends the chosen (hence every) product cone for A and B in C to a product cone for FA and FB in $\mathcal{D}$, and also maps the terminal object of C to a terminal object in $\mathcal{D}$. It is then routine to see that, for cartesian categories C and $\mathcal{D}$, a functor $F : C \to \mathcal{D}$ preserves finite products if and only if it is a strong symmetric monoidal functor (with respect to the symmetric monoidal structure described in the last example). Similarly, we say that a functor between cartesian categories *strictly preserves finite products* if it sends the chosen product cones to the chosen product cones and the chosen terminal object to the chosen terminal object. Then a functor strictly preserves finite products if and only if it is a strict symmetric monoidal functor. Note that there are symmetric monoidal functors between cartesian categories which do not preserve finite products (e.g. the covariant powerset functor on the category of sets). □

Definition 3.1.10 (monoidal natural transformations [30])
For monoidal functors (F, m, m_I), (G, n, n_I) with the same source and target monoidal categories, a *monoidal natural transformation* from (F, m, m_I) to (G, n, n_I) is a natural transformation $\varphi : F \to G$ such that the following diagrams commute:

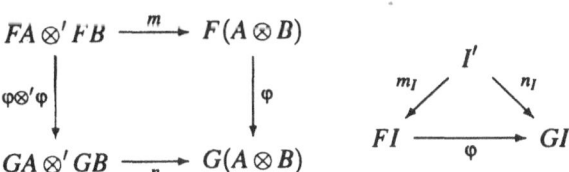

□

In the case of strict functors, a monoidal natural transformation φ between them is a natural transformation satisfying $\varphi_{A \otimes B} = \varphi_A \otimes \varphi_B$ and $\varphi_I = id_I$.

3.2 Acyclic Sharing Models

The correspondence between algebraic theories and finite product preserving functors, originated by Lawvere [58], is now standard – see [26] for a detailed account. We shall briefly recall this story, to help with our descriptions of our story for sharing theories.

Given an algebraic theory $\mathbb{T}$, one can give a model M of the theory as a sound interpretation in a cartesian category C. It is also possible to consider the category of models in C and homomorphisms between them, for which we shall write $\mathbf{Mod}(\mathbb{T}, C)$. By the way, we can construct a cartesian category from $\mathbb{T}$ in syntactic way, which, denoted by $C_{\mathbb{T}}$ and called the *classifying category* or the *term model*, gives an obvious complete model in it. And it gives a pleasant natural equivalence

$$\mathbf{FP}(C_{\mathbb{T}}, C) \simeq \mathbf{Mod}(\mathbb{T}, C)$$

where $\mathbf{FP}$ is the category of small cartesian categories and finite product preserving functors. On the other hand, with any cartesian category C, we can associate an algebraic theory and these correspondences give an equivalence of categories of theories and models.

The departure from this standard story is a simple observation that cartesian categories are not sufficient, or too strong, for interpreting the notion of sharing. We wish to separate the meaning of let (x) be M in N from that of $N\{M/x\}$ if M contains some uncopiable resource (operator symbol), because in $N\{M/x\}$, M can be duplicated or discarded according to the number of x's occurring in N. But this distinction is not possible in the conventional interpretation of algebraic theories in cartesian categories, as both of them are simply interpreted as a composition of arrows corresponding to M and N. Then the natural idea is to replace cartesian categories by symmetric monoidal categories whose axiomatization allows just the linear treatment of resources. However, for modeling sharing graph, we also need to treat variables (pointer-names) in a non-linear way, so we want to keep the benefit of cartesian categories. Therefore we are led to use, instead of cartesian categories, identity-on-objects, strict symmetric monoidal functors from cartesian categories to symmetric monoidal categories. Though not necessarily true, we may intuitively consider this as a symmetric monoidal category with a sub-cartesian category which is full on objects. This amalgam of linear and non-linear settings, which we will call a *cartesian-center symmetric monoidal category*, works well as models of sharing graphs. We interpret linear resources (operator symbols) directly in the symmetric monoidal category, while non-linear variables are first interpreted in the cartesian category part and then imported into the symmetric monoidal category via the strict functor.

Definition 3.2.1 (cartesian-center symmetric monoidal categories)
A *cartesian-center symmetric monoidal category* (*cartesian-center SMC*) is a strict symmetric monoidal functor $\mathcal{F}$ from a cartesian category C to a symmetric monoidal category S which is identity on objects. It is *strict* if both C and S are strict, and *faithful* if $\mathcal{F}$ is faithful (i.e. C is a subcategory of S). □

Remark 3.2.2 The word "center" is taken from work by Power and Robinson [77, 76] on *premonoidal categories*. Informally, a premonoidal category is a monoidal category without bifunctoriality – so, for $f : A \to B$ and $g : A' \to B'$, $(f \otimes A')$; $(B \otimes g)$ and $(A \otimes g)$; $(f \otimes B')$ may not agree in a premonoidal category. An arrow f is called *central* if the equation above holds for any g. Generalizing the classical notion of the center of groups and monoids, Power and Robinson define the subcategory of central maps which is necessarily a monoidal category, and this is the original definition of the center of a premonoidal category. In some interesting applications, a center contains a cartesian subcategory, the inclusion being identity on objects and strict premonoidal, for instance see Thielecke's work on models of continuations [86]. In this thesis we do not need the full generality of Power and Robinson's work, but we prefer to keep this connection, as there seems some natural extension of our work to premonoidal structures with important applications; see the conclusion chapter for further discussion. □

Example 3.2.3 Any cartesian category C is, of course, a cartesian-center symmetric monoidal category, where the symmetric monoidal category S is identical to the cartesian category C, and the functor $\mathcal{F}$ is just the identity on C. □

Example 3.2.4 Typically, one may take C as the category of sets and functions. To model partial computation, S can be the category of sets and partial functions, whose monoidal structure is inherited from the cartesian products of C (which are no longer cartesian in S). To model non-determinism, S can be the category of sets and binary relations, again whose monoidal structure is inherited from the cartesian products of C. Similar examples are available by replacing C by a category of predomains (e.g. ω-cpo's) and S by a suitable category of partial maps etc. Actually many such S arise as the Kleisli categories of commutative monads on C; such cases will play the central role as the semantic models of the higher-order extensions in the next chapter. □

Definition 3.2.5 (cartesian-center functors)
A *cartesian-center symmetric monoidal functor* (shortly *cartesian-center functor*) between cartesian-center SMC's $\mathcal{F} : C \to S$ and $\mathcal{F}' : C' \to S'$ consists of a pair of functors (Φ, Ψ) where $\Phi : C \to C'$ is a strict finite product preserving functor and $\Psi : S \to S'$ is a strict symmetric monoidal functor, satisfying $\mathcal{F}' \circ \Phi = \Psi \circ \mathcal{F}$. □

Definition 3.2.6 (cartesian-center natural transformations)
Let (Φ_1, Ψ_1) and (Φ_2, Ψ_2) be cartesian-center functors from $\mathcal{F} : C \to S$ to $\mathcal{F}' : C' \to S'$. A *cartesian-center monoidal natural transformations* (*cartesian-center natural transformation* for short) from (Φ_1, Ψ_1) to (Φ_2, Ψ_2) consists of a pair of monoidal natural transformations $\alpha : \Phi_1 \dot\to \Phi_2$ and $\beta : \Psi_1 \dot\to \Psi_2$ such that $\mathcal{F}'\alpha = \beta\mathcal{F}$. (Since $\mathcal{F}$ is identity on objects, components of β are determined by α, but we require the naturality in S). □

We write **CcSMC** for the 2-category of small cartesian-center symmetric monoidal categories, cartesian-center functors and cartesian-center natural transformations.

It is well known that cartesian categories satisfy a functional completeness property [56, 41]. We have a parallel result for cartesian-center SMC's as a straightforward generalization (and as a special case of the 2-categorical formulation in [41]):

Lemma 3.2.7 Let $\mathcal{F} : C \to S$ be a cartesian-center SMC and A be an object of C (hence S). Then the Kleisli category $C//A$ of a comonad $A \times (-)$ on C is a cartesian category; the Kleisli category $S//A$ of a comonad $A \otimes (-)$ on S is a symmetric monoidal category; and $\mathcal{F}$ induces an identity-on-objects strict symmetric monoidal functor from $C//A$ to $S//A$ (for which we write $\mathcal{F}//A$). Therefore $\mathcal{F}//A : C//A \to S//A$ is a cartesian-center SMC. $\qquad\square$

The notation $C//A$ is taken from [41] where it is called "the simple slice category over A". $\mathcal{F}//A$ serves as a "polynomial category" [56] in the following sense.

Proposition 3.2.8 (functional completeness)
The obvious identity-on-objects cartesian-center functor $(I, \mathcal{J})$ from $\mathcal{F} : C \to S$ to $\mathcal{F}//A : C//A \to S//A$ satisfies the following universal property. For any identity-on-objects cartesian-center functor (Φ, Ψ) from $\mathcal{F} : C \to S$ to $\mathcal{F}' : C' \to S'$ with an arrow $a : 1 \to A$ in C', there is a unique cartesian-center functor (Φ_a, Ψ_a) from $\mathcal{F}//A : C//A \to S//A$ to $\mathcal{F}' : C' \to S'$ such that $(\Phi_a, \Psi_a) \circ (I, \mathcal{J}) = (\Phi, \Psi)$ and $\Phi_a(\pi_{A,1}) = a$. $\qquad\square$

(In general the universality should be stated up to isomorphism, as done in [41], but here we deal with functors which preserve structure on the nose.)

Though these observations are obvious in this basic setting, they will remain true after introducing additional requirements for interpreting higher-order and cyclic sharing, and turn out to be useful for simplifying our calculation (especially in the cyclic setting, see Lemma 6.2.3 and the proofs of Theorem 7.1.1 and Theorem 7.2.1).

Now let us consider how to interpret sharing theories in our categorical structure. We proceed as follows: define the notion of structure for a signature; give the notion of models; and then show a soundness theorem. A completeness theorem follows after constructing the classifying category in the next section.

Definition 3.2.9 (sharing structures)
A *sharing structure* over an S-sorted signature Σ in a strict cartesian-center SMC $\mathcal{F} : C \to S$ is a pair of functions $[\![-]\!]_S : S \to \mathrm{Obj}(S)$ and $[\![-]\!]_\Sigma : \Sigma \to \mathrm{Arr}(S)$ (subscripts may be omitted, and we just overload $[\![-]\!]$ for both of them) such that $[\![F]\!]$ is an arrow from $[\![(\sigma_1, \ldots, \sigma_m)]\!]$ to $[\![(\tau_1, \ldots, \tau_n)]\!]$ for each operator symbol $F : (\sigma_1, \ldots, \sigma_m) \to (\tau_1, \ldots, \tau_n)$ of Σ, where $[\![(\sigma_1, \ldots, \sigma_m)]\!] = [\![\sigma_1]\!] \otimes \ldots \otimes [\![\sigma_m]\!]$. $\qquad\square$

Given a sharing structure $[\![-]\!]$ in a strict cartesian-center SMC $F : C \to S$, we define $[\![x_1 : \sigma_1, \ldots, x_m : \sigma_m \vdash M : (\tau_1, \ldots, \tau_n)]\!] : [\![(\sigma_1, \ldots, \sigma_m)]\!] \to [\![(\tau_1, \ldots, \tau_n)]\!]$ in S for each well-typed term $x_1 : \sigma_1, \ldots, x_m : \sigma_m \vdash M : (\tau_1, \ldots, \tau_n)$ as follows, by induction on the typing rules.

$$
\begin{aligned}
[\![\Gamma, x : \sigma \vdash x : (\sigma)]\!] &= \mathcal{F}(\pi') \\
[\![\Gamma \vdash F(M) : (\vec{\tau})]\!] &= [\![\Gamma \vdash M : (\vec{\sigma})]\!]; [\![F]\!] \\
[\![\Gamma \vdash 0 : ()]\!] &= \mathcal{F}(!) \\
[\![\Gamma \vdash M \otimes N : (\vec{\sigma}, \vec{\tau})]\!] &= \mathcal{F}(\Delta); ([\![\Gamma \vdash M : (\vec{\sigma})]\!] \otimes [\![\Gamma \vdash N : (\vec{\tau})]\!]) \\
[\![\Gamma \vdash \mathrm{let}\ (\vec{x})\ \mathrm{be}\ M\ \mathrm{in}\ N : (\vec{\tau})]\!] &= \mathcal{F}(\Delta); (id \otimes [\![\Gamma \vdash M : (\vec{\sigma})]\!]); [\![\Gamma, \vec{x} : \vec{\sigma} \vdash N : (\vec{\tau})]\!] \\
[\![\Gamma, x' : \sigma', x : \sigma, \Gamma' \vdash M : (\vec{\tau})]\!] &= (id \otimes \mathcal{F}(c) \otimes id); [\![\Gamma, x : \sigma, x' : \sigma', \Gamma' \vdash M : (\vec{\tau})]\!]
\end{aligned}
$$

where $\Delta_X = \langle id_X, id_X \rangle : X \to X \times X$ in C. Intuitively, this definition corresponds to the inductive constructions of graphs, as shown by pictures below (reproduced from Chapter 2).

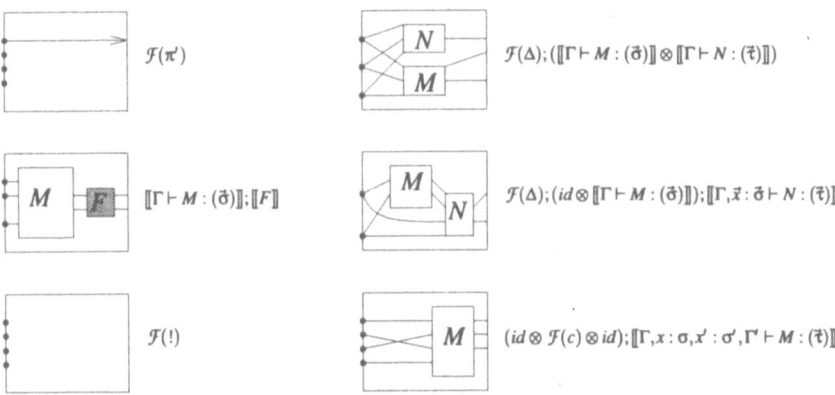

Note that the derivations of variables and let-bindings amount to identity arrows and composition in the simple slice category $S//[\![|\Gamma|]\!]$, where $|\Gamma|$ is defined by

$$|x_1 : \sigma_1, \ldots, x_n : \sigma_n| = (\sigma_1, \ldots, \sigma_n).$$

Remark 3.2.10 Though a term may have many derivations, it is straightforward to show that they all give equal interpretations, and therefore the slightly loose definition above does not cause problems.   □

Lemma 3.2.11 Let $[\![-]\!]$ be a sharing structure. Then $[\![\vec{x} : \vec{\sigma} \vdash M : (\vec{\tau})]\!] = [\![\vec{y} : \vec{\sigma} \vdash M\{\vec{y}/\vec{x}\} : (\vec{\tau})]\!]$ where the x's and y's are disjoint.

Proof: Induction along the construction of M.   □

Definition 3.2.12 (sharing models)
A *sharing model* of an acyclic sharing theory is a sharing structure $[\![-]\!]$ in a cartesian-center SMC $\mathcal{F} : C \to S$ such that $[\![\Gamma \vdash M : (\vec{\sigma})]\!] = [\![\Gamma \vdash N : (\vec{\sigma})]\!]$ for each additional axiom $\Gamma \vdash M = N : (\vec{\sigma})$ of the theory.   □

Therefore a structure is automatically a model of the pure sharing theory (which does not have any additional axiom).

Theorem 3.2.13 (soundness)
Let $[\![-]\!]$ be a sharing model of a sharing theory. If $\Gamma \vdash M = N : (\vec{\sigma})$ is derivable in the theory, then $[\![\Gamma \vdash M : (\vec{\sigma})]\!] = [\![\Gamma \vdash N : (\vec{\sigma})]\!]$.

Proof: We check the axioms in Definition 2.2.3 (page 25).

- (σ_{var}) is proved by induction on the construction of M. If M is x, this is an instance of (id). If M is either another variable or 0, it is easy to see that $[\![\Gamma \vdash$

let (x) be y in $M : (\vec{\sigma})]\!] = [\![\Gamma \vdash M : (\vec{\sigma})]\!]$. For other cases, operators:

$$
\begin{aligned}
&[\![\Gamma \vdash \text{let } (x) \text{ be } y \text{ in } F(M) : (\vec{\tau})]\!] \\
=\ &\mathcal{F}\Delta; (id \otimes [\![\Gamma \vdash y : (\sigma)]\!]); [\![\Gamma, x : \sigma \vdash M]\!]; [\![F]\!] \\
=\ &[\![\Gamma \vdash \text{let } (x) \text{ be } y \text{ in } M : (\vec{\sigma})]\!]; [\![F]\!] \\
=\ &[\![\Gamma \vdash M\{y/x\} : (\vec{\sigma})]\!]; [\![F]\!] \qquad\qquad\qquad \text{induction hypothesis} \\
=\ &[\![\Gamma \vdash F(M)\{y/x\} : (\vec{\tau})]\!]
\end{aligned}
$$

tensor products:

$$
\begin{aligned}
&[\![\Gamma \vdash \text{let } (x) \text{ be } y \text{ in } M \otimes N : (\vec{\tau}, \vec{\tau}')]\!] \\
=\ &\mathcal{F}\Delta; (id \otimes [\![\Gamma \vdash y : (\sigma)]\!]); \mathcal{F}\Delta; \\
&\quad ([\![\Gamma, x : \sigma \vdash M : (\vec{\tau})]\!] \otimes [\![\Gamma, x : \sigma \vdash N : (\vec{\tau}')]\!]) \\
=\ &\mathcal{F}\Delta; (\mathcal{F}\Delta; (id \otimes [\![\Gamma \vdash y : (\sigma)]\!]); [\![\Gamma, x : \sigma \vdash M : (\vec{\tau})]\!] \otimes \\
&\quad \mathcal{F}\Delta; (id \otimes [\![\Gamma \vdash y : (\sigma)]\!]); [\![\Gamma, x : \sigma \vdash N : (\vec{\tau}')]\!]) \\
=\ &\mathcal{F}\Delta; ([\![\Gamma \vdash \text{let } (x) \text{ be } y \text{ in } M : (\vec{\tau})]\!] \otimes [\![\Gamma \vdash \text{let } (x) \text{ be } y \text{ in } N : (\vec{\tau}')]\!]) \\
=\ &\mathcal{F}\Delta; ([\![\Gamma \vdash M\{y/x\} : (\vec{\tau})]\!] \otimes [\![\Gamma \vdash N\{y/x\} : (\vec{\tau}')]\!]) \qquad \text{ind.hyp.} \\
=\ &[\![\Gamma \vdash (M \otimes N)\{y/x\} : (\vec{\tau}, \vec{\tau}')]\!]
\end{aligned}
$$

let-bindings:

$$
\begin{aligned}
&[\![\Gamma \vdash \text{let } (x) \text{ be } y \text{ in let } (\vec{z}) \text{ be } M \text{ in } N : (\vec{\tau})]\!] \\
=\ &\mathcal{F}\Delta; (id \otimes [\![\Gamma \vdash y : (\sigma)]\!]); \mathcal{F}\Delta; \\
&\quad (id \otimes [\![\Gamma, x : \sigma \vdash M : (\vec{\sigma}')]\!]); [\![\Gamma, x : \sigma, \vec{z} : \vec{\sigma}' \vdash N : (\vec{\tau})]\!] \\
=\ &\mathcal{F}\Delta; (id \otimes \mathcal{F}\Delta; (id \otimes [\![\Gamma \vdash y : (\sigma)]\!]); [\![\Gamma, x : \sigma \vdash M : (\vec{\sigma}')]\!]); \\
&\quad \mathcal{F}\Delta; (id \otimes [\![\Gamma, \vec{z} : \vec{\sigma}' \vdash y : (\sigma)]\!]); [\![\Gamma, \vec{z} : \vec{\sigma}', x : \sigma \vdash N : (\vec{\tau})]\!] \\
=\ &\mathcal{F}\Delta; (id \otimes [\![\Gamma \vdash \text{let } (x) \text{ be } y \text{ in } M : (\vec{\sigma}')]\!]); \\
&\quad [\![\Gamma, \vec{z} : \vec{\sigma}' \vdash \text{let } (x) \text{ be } y \text{ in } N : (\vec{\tau})]\!] \\
=\ &\mathcal{F}\Delta; (id \otimes [\![\Gamma \vdash M\{y/x\} : (\vec{\sigma}')]\!]); [\![\Gamma, \vec{z} : \vec{\sigma}' \vdash N\{y/x\} : (\vec{\tau})]\!] \quad \text{ind.hyp.} \\
=\ &[\![\Gamma \vdash \text{let } (\vec{z}) \text{ be } M\{y/x\} \text{ in } N\{y/x\} : (\vec{\tau})]\!]
\end{aligned}
$$

- (id):

$$
\begin{aligned}
&[\![\Gamma \vdash \text{let } \vec{x} \text{ be } M \text{ in } \vec{x} : (\vec{\tau})]\!] \\
=\ &\mathcal{F}\Delta; (id \otimes [\![\Gamma \vdash M : (\vec{\tau})]\!]); [\![\Gamma, \vec{x} : \vec{\tau} \vdash \vec{x} : (\vec{\tau})]\!] \\
=\ &[\![\Gamma \vdash M : (\vec{\tau})]\!]
\end{aligned}
$$

- (ass_1):

$$
\begin{aligned}
&[\![\Gamma \vdash \text{let } (\vec{x}) \text{ be } (\text{let } (\vec{y}) \text{ be } L \text{ in } M) \text{ in } N : (\vec{\tau})]\!] \\
=\ &\mathcal{F}\Delta; (id \otimes [\![\Gamma \vdash \text{let } (\vec{y}) \text{ be } L \text{ in } M : (\vec{\sigma}_1)]\!]); [\![\Gamma, \vec{x} : \vec{\sigma}_1 \vdash N : (\vec{\tau})]\!] \\
=\ &\mathcal{F}\Delta; (id \otimes \mathcal{F}\Delta; (id \otimes [\![\Gamma \vdash L : (\vec{\sigma}_2)]\!]); [\![\Gamma, \vec{y} : \vec{\sigma}_2 \vdash M : (\vec{\sigma}_1)]\!])); \\
&\quad [\![\Gamma, \vec{x} : \vec{\sigma}_1 \vdash N : (\vec{\tau})]\!] \\
=\ &\mathcal{F}\Delta; (id \otimes [\![\Gamma \vdash L : (\vec{\sigma}_2)]\!]); \mathcal{F}\Delta; (id \otimes [\![\Gamma, \vec{y} : \vec{\sigma}_2 \vdash M : (\vec{\sigma}_1)]\!]); \\
&\quad [\![\Gamma, \vec{y} : \vec{\sigma}_2, \vec{x} : \vec{\sigma}_1 \vdash N : (\vec{\tau})]\!] \\
=\ &\mathcal{F}\Delta; (id \otimes [\![\Gamma \vdash L : (\vec{\sigma}_2)]\!]); [\![\Gamma, \vec{y} : \vec{\sigma}_2 \vdash \text{let } (\vec{x}) \text{ be } M \text{ in } N : (\vec{\tau})]\!] \\
=\ &[\![\Gamma \vdash \text{let } (\vec{y}) \text{ be } L \text{ in let } (\vec{x}) \text{ be } M \text{ in } N : (\vec{\tau})]\!]
\end{aligned}
$$

(This amounts to the associativity of compositions in $\mathcal{S}//[\![|\Gamma|]\!]$.)

- (ass$_2$):

$$
\begin{aligned}
&[\![\Gamma \vdash \text{let } (\vec{x}) \text{ be } L \text{ in let } (\vec{y}) \text{ be } M \text{ in } N : (\vec{\tau})]\!] \\
=\ &\mathcal{F}\Delta; (id \otimes [\![\Gamma \vdash L : (\vec{\sigma}_1)]\!]); [\![\Gamma, \vec{x} : \vec{\sigma}_1 \vdash \text{let } (\vec{y}) \text{ be } M \text{ in } N : (\vec{\tau})]\!] \\
=\ &\mathcal{F}\Delta; (id \otimes [\![\Gamma \vdash L : (\vec{\sigma}_1)]\!]); \mathcal{F}\Delta; (id \otimes [\![\Gamma, \vec{x} : \vec{\sigma}_1 \vdash M : (\vec{\sigma}_2)]\!]); \\
&[\![\Gamma, \vec{x} : \vec{\sigma}_1, \vec{y} : \vec{\sigma}_2 \vdash N : (\vec{\tau})]\!] \\
=\ &\mathcal{F}\Delta; (id \otimes \mathcal{F}\Delta; ([\![\Gamma \vdash L : (\vec{\sigma}_1)]\!] \otimes [\![\Gamma \vdash M : (\vec{\sigma}_2)]\!])); \\
&[\![\Gamma, \vec{x} : \vec{\sigma}_1, \vec{y} : \vec{\sigma}_2 \vdash N : (\vec{\tau})]\!] \\
=\ &\mathcal{F}\Delta; (id \otimes [\![\Gamma \vdash L \otimes M : (\vec{\sigma}_1, \vec{\sigma}_2)]\!]); [\![\Gamma, \vec{x} : \vec{\sigma}_1, \vec{y} : \vec{\sigma}_2 \vdash N : (\vec{\tau})]\!] \\
=\ &[\![\Gamma \vdash \text{let } (\vec{x}, \vec{y}) \text{ be } L \otimes M \text{ in } N : (\vec{\tau})]\!]
\end{aligned}
$$

- ($\otimes_1$):

$$
\begin{aligned}
&[\![\Gamma \vdash L \otimes (\text{let } (\vec{x}) \text{ be } M \text{ in } N) : (\vec{\sigma}, \vec{\tau})]\!] \\
=\ &\mathcal{F}\Delta; ([\![\Gamma \vdash L : (\vec{\sigma})]\!] \otimes [\![\Gamma \vdash \text{let } (\vec{x}) \text{ be } M \text{ in } N : (\vec{\tau})]\!]) \\
=\ &\mathcal{F}\Delta; ([\![\Gamma \vdash L : (\vec{\sigma})]\!] \otimes \mathcal{F}\Delta; (id \otimes [\![\Gamma \vdash M : (\vec{\sigma}')]\!]); [\![\Gamma, \vec{x} : \vec{\sigma}' \vdash N : (\vec{\tau})]\!]) \\
=\ &\mathcal{F}\Delta; (id \otimes [\![\Gamma \vdash M : (\vec{\sigma}')]\!]); \mathcal{F}\Delta; \\
&([\![\Gamma, \vec{x} : \vec{\sigma}' \vdash L : (\vec{\sigma})]\!] \otimes [\![\Gamma, \vec{x} : \vec{\sigma}' \vdash N : (\vec{\tau})]\!]) \\
=\ &\mathcal{F}\Delta; (id \otimes [\![\Gamma \vdash M : (\vec{\sigma}')]\!]); [\![\Gamma, \vec{x} : \vec{\sigma}' \vdash L \otimes N : (\vec{\sigma}, \vec{\tau})]\!] \\
=\ &[\![\Gamma \vdash \text{let } (\vec{x}) \text{ be } M \text{ in } L \otimes N : (\vec{\sigma}, \vec{\tau})]\!]
\end{aligned}
$$

- (subst):

$$
\begin{aligned}
&[\![\Gamma \vdash \text{let } (\vec{x}) \text{ be } M \text{ in } F(N) : (\vec{\tau})]\!] \\
=\ &\mathcal{F}\Delta; (id \otimes [\![\Gamma \vdash M : (\vec{\sigma}')]\!]); [\![\Gamma, \vec{x} : \vec{\sigma}' \vdash F(N) : (\vec{\tau})]\!] \\
=\ &\mathcal{F}\Delta; (id \otimes [\![\Gamma \vdash M : (\vec{\sigma}')]\!]); [\![\Gamma, \vec{x} : \vec{\sigma}' \vdash N : (\vec{\sigma})]\!]; [\![F]\!] \\
=\ &[\![\Gamma \vdash \text{let } (\vec{x}) \text{ be } M \text{ in } N : (\vec{\sigma})]\!]; [\![F]\!] \\
=\ &[\![\Gamma \vdash F(\text{let } (\vec{x}) \text{ be } M \text{ in } N) : (\vec{\tau})]\!]
\end{aligned}
$$

$\square$

Definition 3.2.14 (homomorphisms between models)
Let $[\![-]\!]$, $[\![-]\!]'$ be sharing models of a sharing theory in a strict cartesian-center SMC $\mathcal{F} : C \to S$. A *homomorphism* between $[\![-]\!]$ and $[\![-]\!]'$ is a family of arrows $h_\sigma : [\![\sigma]\!] \to [\![\sigma]\!]'$ for each sort σ, such that for each operator symbol $F : (\sigma_1, \ldots, \sigma_m) \to (\tau_1, \ldots, \tau_n)$ the following diagram commutes.

$$
\begin{array}{ccc}
[\![\sigma_1]\!] \otimes \ldots \otimes [\![\sigma_m]\!] & \xrightarrow{\ [\![F]\!]\ } & [\![\tau_1]\!] \otimes \ldots \otimes [\![\tau_n]\!] \\
\Big\downarrow{\scriptstyle h_{\sigma_1} \otimes \ldots \otimes h_{\sigma_m}} & & \Big\downarrow{\scriptstyle h_{\tau_1} \otimes \ldots \otimes h_{\tau_n}} \\
[\![\sigma_1]\!]' \otimes \ldots \otimes [\![\sigma_m]\!]' & \xrightarrow[\ [\![F]\!]'\ ]{} & [\![\tau_1]\!]' \otimes \ldots \otimes [\![\tau_n]\!]'
\end{array}
$$

$\square$

For a sharing theory $\mathbb{T}$ (determined by a set of axioms) over a signature Σ, we write **SharingMod**$(\mathbb{T}, (\mathcal{F} : C \to S))$ for the category of sharing models of $\mathbb{T}$ in a strict cartesian-center SMC $\mathcal{F} : C \to S$ and the homomorphisms between the models.

After introducing the notion of classifying category (term model) $\mathcal{F}_{\mathbb{T}} : C_{\mathbb{T}} \to S_{\mathbb{T}}$ for the theory $\mathbb{T}$, we will show that there is an equivalence of categories

$$\mathbf{CcSMC}((\mathcal{F}_{\mathbb{T}} : C_{\mathbb{T}} \to S_{\mathbb{T}}), (\mathcal{F} : C \to S)) \simeq \mathbf{SharingMod}(\mathbb{T}, (\mathcal{F} : C \to S)).$$

3.3   The Classifying Category

In this section we construct a sharing model from an acyclic sharing theory syntactically. This is done in a similar manner to the standard way to construct a term model from an algebraic theory, see for instance [26].

Proposition 3.3.1 Given an acyclic sharing theory $\mathbb{T}$ over Σ, the following data give rise to a strict cartesian-center symmetric monoidal category $\mathcal{F}_{\mathbb{T}} : C_{\mathbb{T}} \to S_{\mathbb{T}}$.

- Objects are finite lists of sorts.

- An arrow of the cartesian category $C_{\mathbb{T}}$ from $(\vec{\sigma})$ to $(\vec{\tau})$ is a term of the acyclic sharing theory of the form $\vec{x} : \vec{\sigma} \vdash \vec{y} : (\vec{\tau})$. (Note: we work up to renaming of free variables, i.e. we identify $\vec{x} : \vec{\sigma} \vdash M : (\vec{\tau})$ with $\vec{y} : \vec{\sigma} \vdash M\{\vec{y}/\vec{x}\} : (\vec{\tau})$ for fresh variables $\vec{y}$.)

$$id_{(\sigma_1,\ldots,\sigma_m)} = x_1 : \sigma_1, \ldots, x_m : \sigma_m \vdash x_1 \otimes \ldots \otimes x_m : (\sigma_1, \ldots, \sigma_m)$$

$$(\Gamma \vdash \vec{y} : (\vec{\sigma})); (\vec{z} : \vec{\sigma} \vdash N : (\vec{\tau})) = \Gamma \vdash N\{\vec{x}/\vec{y}\} : (\vec{\tau})$$

The terminal object is the empty list $()$, and the product of $(\vec{\sigma})$ and $(\vec{\tau})$ is $(\vec{\sigma}, \vec{\tau})$. Terminal maps, projections and pairings are given by

$$!_{(\vec{\sigma})} = \vec{x} : \vec{\sigma} \vdash 0 : ()$$

$$\pi_{(\vec{\sigma}),(\vec{\tau})} = \vec{x} : \vec{\sigma}, \vec{y} : \vec{\tau} \vdash \vec{x} : (\vec{\sigma})$$

$$\pi'_{(\vec{\sigma}),(\vec{\tau})} = \vec{x} : \vec{\sigma}, \vec{y} : \vec{\tau} \vdash \vec{y} : (\vec{\tau})$$

$$\langle \Gamma \vdash M : (\vec{\sigma}), \Gamma \vdash N : (\vec{\tau}) \rangle = \Gamma \vdash M \otimes N : (\vec{\sigma}, \vec{\tau})$$

- An arrow of the symmetric monoidal category $S_{\mathbb{T}}$ from $(\vec{\sigma})$ to $(\vec{\tau})$ is an equivalence class of terms of the theory $\vec{x} : \vec{\sigma} \vdash M : (\vec{\tau})$, for which we shall write $[\vec{x} : \vec{\sigma} \vdash M : (\vec{\tau})]$. Identities and compositions are

$$id_{(\sigma_1,\ldots,\sigma_m)} = [x_1 : \sigma_1, \ldots, x_m : \sigma_m \vdash x_1 \otimes \ldots \otimes x_m : (\sigma_1, \ldots, \sigma_m)]$$

$$[\Gamma \vdash M : (\vec{\sigma})]; [\vec{x} : \vec{\sigma} \vdash N : (\vec{\tau})] = [\Gamma \vdash \text{let } (\vec{x}) \text{ be } M \text{ in } N : (\vec{\tau})]$$

The unit object is the empty list $()$, and the tensor product of $(\vec{\sigma})$ and $(\vec{\tau})$ is $(\vec{\sigma}, \vec{\tau})$. Tensor products of arrows are given by

$$[\Gamma \vdash M : (\vec{\sigma})] \otimes [\Gamma' \vdash N : (\vec{\tau})] = [\Gamma, \Gamma' \vdash M \otimes N : (\vec{\sigma}, \vec{\tau})]$$

where variables in Γ and in Γ' are disjoint.

- $\mathcal{F}_T$ maps C_T's arrow $\vec{x} : \vec{\sigma} \vdash y_1 \otimes \ldots \otimes y_n : (\vec{\tau})$ to $\mathcal{S}_T$'s arrow $[\vec{x} : \vec{\sigma} \vdash y_1 \otimes \ldots \otimes y_n : (\vec{\tau})]$.

<u>Proof:</u> To check that C_T forms a strict cartesian category is routine and easy – see also the remark below. We shall see that $\mathcal{S}_T$ is a strict symmetric monoidal category, by checking that the axioms for strict symmetric monoidal categories (page 41) are satisfied.

- Identity laws (M1):

$$
\begin{aligned}
& [\Gamma \vdash M : (\vec{\sigma})] ; [\vec{x} : \vec{\sigma} \vdash \vec{x} : (\vec{\sigma})] \\
=\ & [\Gamma \vdash \text{let } (\vec{x}) \text{ be } M \text{ in } \vec{x} : (\vec{\sigma})] \\
=\ & [\Gamma \vdash M : (\vec{\sigma})] \qquad\qquad\qquad\quad \text{(id)}
\end{aligned}
$$

$$
\begin{aligned}
& [\vec{x} : \vec{\sigma} \vdash \vec{x} : (\vec{\sigma})] ; [\vec{y} : \vec{\sigma} \vdash M : (\vec{\tau})] \\
=\ & [\vec{x} : \vec{\sigma} \vdash \text{let } (\vec{y}) \text{ be } \vec{x} \text{ in } M : (\vec{\tau})] \\
=\ & [\vec{x} : \vec{\sigma} \vdash M\{\vec{x}/\vec{y}\} : (\vec{\tau})] \qquad\quad (\sigma_{\text{var}}) \\
=\ & [\vec{y} : \vec{\sigma} \vdash M : (\vec{\tau})]
\end{aligned}
$$

- Unit law of tensor products (M2):

$$
[\Gamma \vdash M : (\vec{\sigma})] \otimes [\vdash 0 : ()] = [\Gamma \vdash M \otimes 0 : (\vec{\sigma})] = [\Gamma \vdash M : (\vec{\sigma})]
$$

The other one is similar.

- tensor products of identities (M3):

$$
[\vec{x} : \vec{\sigma} \vdash \vec{x} : (\vec{\sigma})] \otimes [\vec{y} : \vec{\tau} \vdash \vec{y} : (\vec{\tau})] = [\vec{x} : \vec{\sigma}, \vec{y} : \vec{\tau} \vdash \vec{x} \otimes \vec{y} : (\vec{\sigma}, \vec{\tau})]
$$

- Associativity of compositions (M4):

$$
\begin{aligned}
& ([\Gamma \vdash L : (\vec{\sigma})] ; [\vec{x} : \vec{\sigma} \vdash M : (\vec{\sigma}')]) ; [\vec{y} : \vec{\sigma}' \vdash L : (\vec{\tau})] \\
=\ & [\Gamma \vdash \text{let } (\vec{x}) \text{ be } L \text{ in } M : (\vec{\sigma}')] ; [\vec{y} : \vec{\sigma}' \vdash L : (\vec{\tau})] \\
=\ & [\Gamma \vdash \text{let } (\vec{y}) \text{ be } (\text{let } (\vec{x}) \text{ be } L \text{ in } M) \text{ in } N : (\vec{\tau})] \\
=\ & [\Gamma \vdash \text{let } (\vec{x}) \text{ be } L \text{ in let } (\vec{y}) \text{ be } M \text{ in } N : (\vec{\tau})] \qquad (\text{ass}_1) \\
=\ & [\Gamma \vdash L : (\vec{\sigma})] ; [\vec{x} : \vec{\sigma} \vdash \text{let } (\vec{y}) \text{ be } M \text{ in } N : (\vec{\tau})] \\
=\ & [\Gamma \vdash L : (\vec{\sigma})] ; ([\vec{x} : \vec{\sigma} \vdash M : (\vec{\sigma}')] ; [\vec{y} : \vec{\sigma}' \vdash L : (\vec{\tau})])
\end{aligned}
$$

- Strict associativity of tensor products (M5):

$$
\begin{aligned}
& [\Gamma \vdash L : (\vec{\sigma})] \otimes ([\Gamma' \vdash M : (\vec{\sigma}')] \otimes [\Gamma'' \vdash N : (\vec{\sigma}'')]) \\
=\ & [\Gamma \vdash L : (\vec{\sigma})] \otimes [\Gamma', \Gamma'' \vdash M \otimes N : (\vec{\sigma}', \vec{\sigma}'')] \\
=\ & [\Gamma, \Gamma', \Gamma'' \vdash L \otimes M \otimes N : (\vec{\sigma}, \vec{\sigma}', \vec{\sigma}'')] \\
=\ & [\Gamma, \Gamma' \vdash L \otimes M : (\vec{\sigma}, \vec{\sigma}')] \otimes [\Gamma'' \vdash N : (\vec{\sigma}'')] \\
=\ & ([\Gamma \vdash L : (\vec{\sigma})] \otimes [\Gamma' \vdash M : (\vec{\sigma}')]) \otimes [\Gamma'' \vdash N : (\vec{\sigma}'')]
\end{aligned}
$$

- Interchange law (M6):

$$([\Gamma \vdash M : (\vec{\sigma})]; [\vec{x} : \vec{\sigma} \vdash N : (\vec{\tau})]) \otimes ([\Gamma' \vdash M' : (\vec{\sigma}')]; [\vec{x}' : \vec{\sigma}' \vdash N' : (\vec{\tau}')])$$
$$= [\Gamma \vdash \text{let } (\vec{x}) \text{ be } M \text{ in } N : (\vec{\tau})] \otimes [\Gamma' \vdash \text{let } (\vec{x}') \text{ be } M' \text{ in } N' : (\vec{\tau}')]$$
$$= [\Gamma, \Gamma' \vdash (\text{let } (\vec{x}) \text{ be } M \text{ in } N) \otimes (\text{let } (\vec{x}') \text{ be } M' \text{ in } N') : (\vec{\tau}, \vec{\tau}')]$$
$$= [\Gamma, \Gamma' \vdash \text{let } (\vec{x}) \text{ be } M \text{ in } (N \otimes (\text{let } (\vec{x}') \text{ be } M' \text{ in } N')) : (\vec{\tau}, \vec{\tau}')] \quad (\otimes_2)$$
$$= [\Gamma, \Gamma' \vdash \text{let } (\vec{x}) \text{ be } M \text{ in let } (\vec{x}') \text{ be } M' \text{ in } N \otimes N' : (\vec{\tau}, \vec{\tau}')] \quad (\otimes_1)$$
$$= [\Gamma, \Gamma' \vdash \text{let } (\vec{x}, \vec{x}') \text{ be } M \otimes M' \text{ in } N \otimes N' : (\vec{\tau}, \vec{\tau}')] \quad (\text{ass}_2)$$
$$= [\Gamma, \Gamma' \vdash M \otimes M' : (\vec{\sigma}, \vec{\sigma}')]; [\vec{x} : \vec{\sigma}, \vec{x}' : \vec{\sigma}' \vdash N \otimes N' : (\vec{\tau}, \vec{\tau}')]$$
$$= ([\Gamma \vdash M : (\vec{\sigma})] \otimes [\Gamma' \vdash M' : (\vec{\sigma}')]); ([\vec{x} : \vec{\sigma} \vdash N : (\vec{\tau})] \otimes [\vec{x}' : \vec{\sigma}' \vdash N' : (\vec{\tau}')])$$

- Naturality of symmetry (S1):

$$([\vec{x} : \vec{\sigma} \vdash M : (\vec{\tau})] \otimes [\vec{y} : \vec{\sigma}' \vdash N : (\vec{\tau}')]); [\vec{u} : \vec{\tau}, \vec{v} : \vec{\tau}' \vdash \vec{v} \otimes \vec{u} : (\vec{\tau}', \vec{\tau})]$$
$$= [\vec{x} : \vec{\sigma}, \vec{y} : \vec{\sigma}' \vdash M \otimes N : (\vec{\tau}, \vec{\tau}')]; [\vec{u} : \vec{\tau}, \vec{v} : \vec{\tau}' \vdash \vec{v} \otimes \vec{u} : (\vec{\tau}', \vec{\tau})]$$
$$= [\vec{x} : \vec{\sigma}, \vec{y} : \vec{\sigma}' \vdash \text{let } (\vec{u}, \vec{v}) \text{ be } M \otimes N \text{ in } \vec{v} \otimes \vec{u} : (\vec{\tau}', \vec{\tau})]$$
$$= [\vec{x} : \vec{\sigma}, \vec{y} : \vec{\sigma}' \vdash \text{let } (\vec{u}) \text{ be } M \text{ in let } (\vec{v}) \text{ be } N \text{ in } \vec{v} \otimes \vec{u} : (\vec{\tau}', \vec{\tau})] \quad (\text{ass}_2)$$
$$= [\vec{x} : \vec{\sigma}, \vec{y} : \vec{\sigma}' \vdash \text{let } (\vec{u}) \text{ be } M \text{ in } ((\text{let } (\vec{v}) \text{ be } N \text{ in } \vec{v}) \otimes \vec{u}) : (\vec{\tau}', \vec{\tau})] \quad (\otimes_2)$$
$$= [\vec{x} : \vec{\sigma}, \vec{y} : \vec{\sigma}' \vdash \text{let } (\vec{u}) \text{ be } M \text{ in } N \otimes \vec{u} : (\vec{\tau}', \vec{\tau})] \quad (\text{id})$$
$$= [\vec{x} : \vec{\sigma}, \vec{y} : \vec{\sigma}' \vdash N \otimes (\text{let } (\vec{u}) \text{ be } M \text{ in } \vec{u}) : (\vec{\tau}', \vec{\tau})] \quad (\otimes_1)$$
$$= [\vec{x} : \vec{\sigma}, \vec{y} : \vec{\sigma}' \vdash N \otimes M : (\vec{\tau}', \vec{\tau})] \quad (\text{id})$$
$$= [\vec{x} : \vec{\sigma}, \vec{y} : \vec{\sigma}' \vdash \text{let } (\vec{z}, \vec{w}) \text{ be } \vec{y} \otimes \vec{x} \text{ in } N\{\vec{z}/\vec{y}\} \otimes M\{\vec{w}/\vec{x}\} : (\vec{\tau}', \vec{\tau})] \quad (\sigma_{\text{var}})$$
$$= [\vec{x} : \vec{\sigma}, \vec{y} : \vec{\sigma}' \vdash \vec{y} \otimes \vec{x} : (\vec{\sigma}', \vec{\sigma})]; [\vec{z} : \vec{\sigma}', \vec{w} : \vec{\sigma} \vdash N\{\vec{z}/\vec{y}\} \otimes M\{\vec{w}/\vec{x}\} : (\vec{\tau}', \vec{\tau})]$$
$$= [\vec{x} : \vec{\sigma}, \vec{y} : \vec{\sigma}' \vdash \vec{y} \otimes \vec{x} : (\vec{\sigma}', \vec{\sigma})]; [\vec{y} : \vec{\sigma}', \vec{x} : \vec{\sigma} \vdash N \otimes M : (\vec{\tau}', \vec{\tau})]$$
$$= [\vec{x} : \vec{\sigma}, \vec{y} : \vec{\sigma}' \vdash \vec{y} \otimes \vec{x} : (\vec{\sigma}', \vec{\sigma})]; ([\vec{y} : \vec{\sigma}' \vdash N : (\vec{\tau}')] \otimes [\vec{x} : \vec{\sigma} \vdash M : (\vec{\tau})])$$

S2 and S3 follow from the calculation that $C_{\mathbb{T}}$ is a cartesian category, and then $\mathcal{F}_{\mathbb{T}}$ is obviously strict symmetric monoidal. $\qquad\square$

Remark 3.3.2 $C_{\mathbb{T}}$ is equivalent to the free strict cartesian category whose objects are generated from the set of sorts S. Explicitly, an arrow from $(\sigma_1, \ldots, \sigma_m)$ to $(\tau_1, \ldots, \tau_n)$ is determined by a function f from $\{1, \ldots, n\}$ to $\{1, \ldots, m\}$ such that $\tau_i = \sigma_{f(i)}$; composition is then determined by that of functions. Thus $C_{\mathbb{T}}$ does not depend on the choice of theory $\mathbb{T}$. On the other hand, the symmetric monoidal category part $S_{\mathbb{T}}$ is determined by the elements of Σ as well as the axioms of $\mathbb{T}$. Note that $\mathcal{F}_{\mathbb{T}}$ may not be an inclusion, for instance the axioms of the theory may enforce two different variables to be equal. $\qquad\square$

We say a model is *complete* if $[\![\Gamma \vdash M : (\vec{\tau})]\!] = [\![\Gamma \vdash N : (\vec{\tau})]\!]$ implies $\Gamma \vdash M = N : (\vec{\tau})$ in the theory.

Theorem 3.3.3 (completeness)
Given a theory $\mathbb{T}$, there is a complete model in $\mathcal{F}_{\mathbb{T}} : C_{\mathbb{T}} \to S_{\mathbb{T}}$, given by $[\![\sigma]\!] = \sigma$ and $[\![F]\!] = [\vec{x} : \vec{\sigma} \vdash F(\vec{x}) : (\vec{\tau})]$ for $F : (\vec{\sigma}) \to (\vec{\tau})$. $\qquad\square$

Note that this model is not just complete but also *full*, in that any arrow is the image of a term.

Remark 3.3.4 The reader who followed the details of the proof may notice that the axiom (subst) is not used for showing that the classifying category is a cartesian-center SMC. It is not needed even to show completeness – but needed for fullness; without (subst) the resulting category contains some "junk" morphisms (which can be ignored to show just completeness).

In Chapter 8 we will extend the definition of operators so that they can have parameters for which arbitrary substitution may not be justified, and there these junks will get a proper interpretation and explanation. □

We call $\mathcal{F}_{\mathbb{T}} : C_{\mathbb{T}} \to S_{\mathbb{T}}$ the *classifying category* of the acyclic sharing theory $\mathbb{T}$, and the complete model described above the *generic model* $[\![-]\!]_G$. The names are justified by the following observation.

Lemma 3.3.5 Let $[\![-]\!]$ be a model of $\mathbb{T}$ in a strict cartesian-center SMC $\mathcal{F} : C \to S$, and consider a cartesian-center functor $(\Phi, \Psi) : (\mathcal{F} : C \to S) \to (\mathcal{F}' : C' \to S')$. We shall define a structure $[\![-]\!]'$ in $\mathcal{F}' : C' \to S'$ by $[\![\sigma]\!]' = \Psi([\![\sigma]\!])$ for each sort σ and $[\![F]\!]' = \Psi([\![F]\!])$ for each operator symbol F. Then $[\![-]\!]'$ is a model of $\mathbb{T}$.

<u>Proof:</u> It suffices to show that $[\![\Gamma \vdash M]\!]' = \Psi([\![\Gamma \vdash M]\!])$ by induction on the construction of M. □

Proposition 3.3.6 For any model $[\![-]\!]$ of $\mathbb{T}$ in a strict cartesian-center SMC $\mathcal{F} : C \to S$, there is a unique cartesian-center functor (Φ, Ψ) from the classifying category to $\mathcal{F} : C \to S$ such that $[\![-]\!] = \Psi([\![-]\!]_G)$. Pictorially:

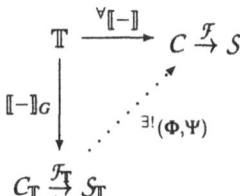

<u>Proof:</u> Since $C_{\mathbb{T}}$ is the free cartesian category, $\Phi : C_{\mathbb{T}} \to C$ is uniquely determined. Then Ψ is required to satisfy

$$\Psi([\![\Gamma, x : \sigma \vdash x : (\sigma)]\!]) = \mathcal{F}(\Phi(\Gamma, x : \sigma \vdash x : (\sigma)))$$
$$\Psi([\![\Gamma \vdash \mathcal{F}(M) : (\vec{\tau})]\!]) = \Psi([\![\Gamma \vdash M : (\vec{\sigma})]\!]) ; [\![F]\!]$$
$$\Psi([\![\Gamma \vdash 0 : ()]\!]) = \mathcal{F}(\Phi(\Gamma \vdash 0 : ()))$$
$$\Psi([\![\Gamma \vdash \text{let } (\vec{x}) \text{ be } M \text{ in } N : (\vec{\tau})]\!]) = \mathcal{F}(\Delta) ; (id \otimes \Psi([\![\Gamma \vdash M : (\vec{\sigma})]\!])) ;$$
$$\Psi([\![\Gamma, \vec{x} : \vec{\sigma} \vdash N : (\vec{\tau})]\!])$$
$$\Psi([\![\Gamma, x : \sigma, x' : \sigma', \Gamma' \vdash M : (\vec{\tau})]\!]) = (id \otimes \mathcal{F}(c) \otimes id) ;$$
$$\Psi([\![\Gamma, x' : \sigma', x : \sigma, \Gamma' \vdash M : (\vec{\tau})]\!])$$

Such a Ψ exists as $[\![-]\!]$ is a model, and is uniquely determined by these equations. □

It is easily seen that, in this setting, there is a natural bijection between the cartesian-center natural transformations and the homomorphisms, and thus we have

Theorem 3.3.7

$$\mathbf{CcSMC}((C_{\mathbb{T}} \overset{\mathcal{F}_{\mathbb{T}}}{\to} S_{\mathbb{T}}), (C \overset{\mathcal{F}}{\to} S)) \simeq \mathbf{SharingMod}(\mathbb{T}, (C \overset{\mathcal{F}}{\to} S)).$$

□

Remark 3.3.8 In the pure case, there is yet another way to construct the classifying category – from sharing graphs. We omit the fairly routine construction; the resulting cartesian-center SMC is then isomorphic to the classifying category obtained as above, and the correspondence between sharing graphs and pure acyclic sharing theory gives the isomorphism of cartesian-center SMC's. This observation provides an "algebraic" justification of Theorem 2.2.16 in the last chapter; as claimed there, the correspondence is not just a bijection but actually a structure preserving isomorphism.

□

3.4 Theory-Model Correspondence

This section is to complete the analogy with the standard categorical type theory. There is no technical significance, but it may be of some conceptual interest for those familiar with the standard theory. A small gap between the traditional *theory-model correspondence* and ours is pointed out.

Let $\mathcal{F} : C \to S$ be a small strict cartesian-center SMC. From this, we construct an acyclic sharing theory $\mathbb{T}_{\mathcal{F}}$ as follows. For simplicity, we assume that objects are freely generated from a set S. First, sorts are given by the set S. The signature of $\mathbb{T}_{\mathcal{F}}$ is then the set of S's arrows, i.e. for each $f : \sigma_1 \otimes \ldots \otimes \sigma_n \to \tau_1 \otimes \ldots \otimes \tau_n$ in S we associate an operator symbol of type $((\sigma_1, \ldots, \sigma_m), (\tau_1, \ldots, \tau_n))$. Then $\mathbb{T}_{\mathcal{F}}$ is an acyclic sharing theory over this signature, with axioms $\Gamma \vdash M = N : (\vec{\tau})$ for all $[\![\Gamma \vdash M : (\vec{\tau})]\!] = [\![\Gamma \vdash N : (\vec{\tau})]\!]$ in S, where $[\![-]\!]$ is determined by the obvious structure in $\mathcal{F} : C \to S$. By definition this $[\![-]\!]$ is a model of $\mathbb{T}_{\mathcal{F}}$ in $\mathcal{F} : C \to S$.

In the standard theory, the classifying category of the algebraic theory obtained from a cartesian category C is equivalent to the original C. In the case of sharing categories, this is not true, because the cartesian category part of a classifying category is always a free cartesian category, while not all cartesian-center SMC's are in this form. The resulting classifying category from $\mathcal{F} : C \to S$ is not $\mathcal{F} : C \to S$ but the composition $C_0 \to C \overset{\mathcal{F}}{\to} S$ where C_0 is the free cartesian category generated from S.

If we talk about this restricted class of cartesian-center SMC's, i.e. those whose domain cartesian category is free, then we will obtain the standard theory-model correspondence. The full subcategory of the restricted models share the same initial object (actually the inclusion has a right adjoint which maps a cartesian-center SMC $\mathcal{F} : C \to S$ to the composition $C_0 \to C \overset{\mathcal{F}}{\to} S$), hence all results developed in this chapter apply equally to the restricted version. However we have chosen the larger class of models for the following reasons:

- Most of cartesian categories arising from the natural semantic models of computation (or mathematics) are of course not free. Typically we want to choose C as the category of sets, as well as category of predomains etc.

- In later chapters, we assume additional structure on cartesian-center SMC's to model higher-order computation. However, none of the restricted version of cartesian-center SMC's fits this new requirement.

In fact, such a restricted class has already been studied by Power, under the name of *elementary control structures* [75], in the context of the models of action calculi (where Power also considers how to deal with parameterized operator symbols in action calculi – we will review this issue in a later chapter on action calculi).

3.5 Modeling Rewriting via Local Preorders

In Chapter 2 we defined the notion of rewriting systems on sharing theories. Following [75] we give the semantic interpretation of rewriting in sharing models.

Definition 3.5.1 A locally small strict symmetric monoidal category $\mathcal{M}$ is *locally preordered* if each homset $\mathcal{M}(X, Y)$ is equipped with a preorder $\rightsquigarrow_{X,Y}$ (we write $f \rightsquigarrow g : X \to Y$ for $f \rightsquigarrow_{X,Y} g$), and they are preserved by compositions and tensor products:

$$\frac{f \rightsquigarrow f' : X \to Y \quad g \rightsquigarrow g' : Y \to Z}{f; g \rightsquigarrow f'; g' : X \to Z}$$

$$\frac{f \rightsquigarrow f' : X \to X' \quad g \rightsquigarrow g' : Y \to Y'}{f \otimes g \rightsquigarrow f' \otimes g' : X \otimes Y \to X' \otimes Y'}$$

□

Proposition 3.5.2 There is a bijection between rewriting systems on an acyclic sharing theory $\mathbb{T}$ and local preorders on $\mathcal{S}_{\mathbb{T}}$ of the classifying category $\mathcal{F}_{\mathbb{T}} : C_{\mathbb{T}} \to \mathcal{S}_{\mathbb{T}}$. □

Definition 3.5.3 (minimality condition)
Let $\mathcal{F} : C \to S$ be a small cartesian-center strict SMC. A local-preorder $\rightsquigarrow$ on S satisfies the *minimality condition* if the images of $\mathcal{F}$ are minimal, i.e. $\mathcal{F}(g) \rightsquigarrow f$ implies $f = \mathcal{F}(g)$.

□

Proposition 3.5.4 There is a bijection between rewriting systems with the minimality condition on an acyclic sharing theory $\mathbb{T}$ and local preorders with the minimality condition on $\mathcal{S}_{\mathbb{T}}$ of the classifying category $\mathcal{F}_{\mathbb{T}} : C_{\mathbb{T}} \to \mathcal{S}_{\mathbb{T}}$.

□

Example 3.5.5 Let C be (the strict equivalent of) the category of sets and functions, S be (the strict equivalent of) the category of sets and total relations[1] and $\mathcal{F}$ be the obvious identity-on-objects inclusion functor from C to S, which strictly maps the cartesian products of C to the symmetric monoidal products of S (both given by the direct products of sets). We give a local preorder $\rightsquigarrow$ on S by the inclusion of relations, i.e. $R \rightsquigarrow R'$ iff $R \supseteq R'$. Then $\rightsquigarrow$ satisfies the minimality condition – a total relation is minimal w.r.t. $\rightsquigarrow$ if and only if it is a function. We may interpret a non-deterministic

---

[1] A relation $R \subseteq X \times Y$ is called *total* if for any $x \in X$ there is a $y \in Y$ such that xRy.

programming language in this setting. For instance, the non-deterministic primitive zero_or_one in Chapter 1 (page 3), together with rewriting rules zero_or_one ≻ zero and zero_or_one ≻ one, can be interpreted as $[\![\text{zero\_or\_one}]\!] = \{0, 1\} \subseteq \mathbf{N}$, which satisfies $\{0, 1\} \rightsquigarrow \{0\} = [\![\text{zero}]\!]$ and $\{0, 1\} \rightsquigarrow \{1\} = [\![\text{one}]\!]$ as we expect. And we have (with an adequate interpretation of $+$)

$$[\![\ \vdash \text{zero\_or\_one} + \text{zero\_or\_one} : (\text{nat})\ ]\!] \quad = \quad \{0, 1, 2\}$$
$$[\![\ \vdash \text{let } (x) \text{ be zero\_or\_one in } x + x : (\text{nat})\ ]\!] \quad = \quad \{0, 2\}$$

as intuitively explained in Chapter 1.   □

4
Higher-Order Extension

In this chapter we develop a higher-order extension of acyclic sharing theories, i.e. an extension with lambda abstractions and applications. There has been considerable research on such settings, notably higher-order graph rewriting theory, also called lambda graph rewriting systems, which are led by a practical demand for efficient implementation techniques for functional programming languages.

In some sense, it is fairly routine to enrich our theory with lambda terms. As in the standard type theory, we just add term construction rules for lambda abstractions and applications:

$$\frac{\Gamma, \vec{x} : \vec{\sigma} \vdash M : (\vec{\tau})}{\Gamma \vdash \lambda(\vec{x}).M : ((\vec{\sigma}) \Rightarrow (\vec{\tau}))} \qquad \frac{\Gamma \vdash M : ((\vec{\sigma}) \Rightarrow (\vec{\tau})) \quad \Gamma \vdash N : (\vec{\sigma})}{\Gamma \vdash MN : (\vec{\tau})}$$

The main problem is to find the *right* axiomatization for them. It easily turns out that assuming either the full β axiom or η axiom is sufficient for validating any substitutions in the theory, thus we lose the notion of sharing. Actually there is no practical justification for assuming such strong axioms. In practice, substitutions of function closures (lambda abstractions) are acceptable as they can be seen as finished computation (at least with no side effects) and copying them is in general harmless.

Fortunately, this restriction on substitutions to values (variables and lambda abstractions) matches our theory of sharing very well. In fact, later we will see that adding such higher-order constructs is a conservative extension over the original first-order sharing theory.

Remark 4.0.1 There are further possible choices in designing such calculi, which we will not take in this thesis. First, in practical implementations, all we need is the notion of *weak reduction* in which reduction under lambda abstraction and substitution into lambda abstraction are not performed. This is what real practical interpreters do, but since we are developing the syntax and semantics for (equational) reasoning about such implementation issues, we think it natural that our equational theory proves more than the interpreters perform.

Second, there is an issue of *garbage collection* (c.f. Example 2.2.10). As we have already noticed, in dealing with sharing graphs, there can be isolated resources which cannot be accessed from other part of graphs. In many implementations of functional languages, they can be eliminated safely because such junks do not perform any side effects in future (this is not the case for concurrent languages which are full of side-effect like interactions between independent resources!). So one may wish to add axioms for eliminating isolated resources. However, we do not include such axioms a priori – they can be added to strengthen the theory and to restrict the class of models for such specific purposes (e.g. the study of implementations of functional languages), but we wish to keep our approach as general as possible. For instance, we will see that our theory perfectly covers Milner's treatment of higher-order interactive systems as *higher-order action calculi* (Chapter 8); and our models cover a properly wider range of semantic models of computation than those for pure functional computation. We just choose the axioms we need, and we believe this results in a better understanding of the computation we are talking about.   □

4.1 Higher-Order Acyclic Sharing Theory

Definition 4.1.1 (higher-order sorts)
Given a set of sorts S, we define the set of *higher-order sorts* S^H by

$$S^H \ni s,t,\ldots \quad ::= \quad \sigma \mid m \Rightarrow n$$

where σ ranges over S and m, n over finite lists of elements of S^H.   □

Definition 4.1.2 (higher-order acyclic sharing theory)
A *higher-order acyclic sharing theory* over Σ is an equational theory on the well-typed terms with sorts S^H closed under the term construction of acyclic sharing theories plus

$$\frac{\Gamma,\vec{x}:\vec{\sigma} \vdash M : (\vec{t})}{\Gamma \vdash \lambda(\vec{x}).M : ((\vec{\sigma}) \Rightarrow (\vec{t}))} \text{ abstraction}$$

$$\frac{\Gamma \vdash M : ((\vec{\sigma}) \Rightarrow (\vec{t})) \quad \Gamma \vdash N : (\vec{\sigma})}{\Gamma \vdash MN : (\vec{t})} \text{ application}$$

where the equality on terms is a congruence relation containing the axioms for the acyclic sharing theory plus the following additional axioms.

(β)	$(\lambda(\vec{x}).M)N$	=	let $(\vec{x})$ be N in M
(η_0)	$\lambda(\vec{x}).y(\vec{x})$	=	y
(σ_v)	let (x) be $\lambda(\vec{y}).M$ in N	=	$N\{\lambda(\vec{y}).M/x\}$
(app_1)	$(\text{let } (\vec{x}) \text{ be } L \text{ in } M)N$	=	let $(\vec{x})$ be L in MN
(app_2)	$L(\text{let } (\vec{x}) \text{ be } M \text{ in } N)$	=	let $(\vec{x})$ be M in LN

Note that the congruence must be closed under the new term constructions (lambda abstractions and applications):

$$\frac{\Gamma \vdash M = M' : ((\vec{\sigma} \Rightarrow (\vec{t})) \quad \Gamma \vdash N = N' : (\vec{\sigma})}{\Gamma \vdash MN = M'N' : (\vec{t})} \qquad \frac{\Gamma \vdash M = N : (\vec{t})}{\Gamma \vdash \lambda(\vec{x}).M = \lambda(\vec{x}).N : ((\vec{\sigma}) \Rightarrow (\vec{t}))}$$

□

Definition 4.1.3 (values)
Values are well-typed terms generated by the following grammar.

$$V,W ::= 0 \mid x \mid \lambda(\vec{x}).M \mid V \otimes W$$

□

The "call-by-value" $\beta\eta$ equations as well as the α-conversion of lambda-bound variables follow immediately from our axioms:

Lemma 4.1.4 The following equations are derivable in higher-order acyclic sharing theories.

$$
\begin{array}{llll}
(\beta_v) & (\lambda(\vec{x}).M)V & = & M\{V/\vec{x}\} \\
(\eta_v) & (\lambda(\vec{x}).V(\vec{x})) & = & V & (\vec{x} \notin FV(V)) \\
(\alpha) & \lambda(\vec{x}).M & = & \lambda(\vec{y}).M\{\vec{y}/\vec{x}\} & (\vec{y} \text{ are fresh})
\end{array}
$$

$\square$

4.2 Higher-Order Acyclic Sharing Models

Since the higher-order acyclic sharing theories are obtained by adding new terms and axioms to the first-order ones, their models should be a subclass of those for first-order theories. We add a new requirement to cartesian-center SMC's:

Definition 4.2.1 (cartesian centrally closed SMC)
A *cartesian centrally closed symmetric monoidal category* (*cartesian centrally closed SMC*) is a cartesian-center symmetric monoidal category $\mathcal{F} : C \to S$ such that, for each object X of S, the functor $\mathcal{F}(-) \otimes X : C \to S$ has a right adjoint. $\square$

We write $X \Rightarrow (-) : S \to C$ for a chosen right adjoint of $\mathcal{F}(-) \otimes X$, and use **ap** : $(X \Rightarrow A) \otimes X \to A$ in S for the counit of adjunction, as well as $f^* : A \to X \Rightarrow B$ in C for the adjunct of $f : A \otimes X \to B$ in S. However, we often omit the information of the chosen right adjoints if there is no confusion. We note that $\mathcal{F}$ itself has a right adjoint $I \Rightarrow (-)$.

Remark 4.2.2 The word "centrally closed" is taken from the work by Power [76]. In his more general framework, we are talking about special instances of centrally closed premonoidal $\to$-categories. $\square$

Example 4.2.3 Any cartesian closed category can be seen as a cartesian centrally closed symmetric monoidal category. $\square$

Example 4.2.4 Let C be a cartesian closed category and T be a commutative monad (see Definition 5.1.3) on C. Putting $S = C_T$ (the Kleisli category of T) and letting $\mathcal{F}$ be the canonical identity-on-object functor from C to S, we obtain a cartesian centrally closed SMC, where the right adjoint $X \Rightarrow (-)$ is given by $(T-)^X$. The relation with this monad-based account will be further spelled out in the next chapter. $\square$

Example 4.2.5 As an instance of the setting described above, we shall consider the case where C is the category of sets and functions and T is the covariant powerset functor. In this case, $S = C_T$ is the category of sets and binary relations (i.e. non-deterministic functions), and $X \Rightarrow Y$ is simply the set of all relations between X and Y. $\square$

Definition 4.2.6 (cartesian centrally closed functors)

A *cartesian centrally closed functor* between cartesian centrally closed SMC's $\mathcal{F}$: $C \to S$ and $\mathcal{F}' : C' \to S'$ is a cartesian-center functor (Φ, Ψ) between them such that (Φ, Ψ) is a map of adjoints from $\mathcal{F}(-) \otimes X \dashv X \Rightarrow (-)$ to $\mathcal{F}'(-) \otimes' \Psi X \dashv \Psi X \Rightarrow' (-)$ for each object X, i.e. the following diagrams commute (there are many equivalent formulations of this condition – see [59]).

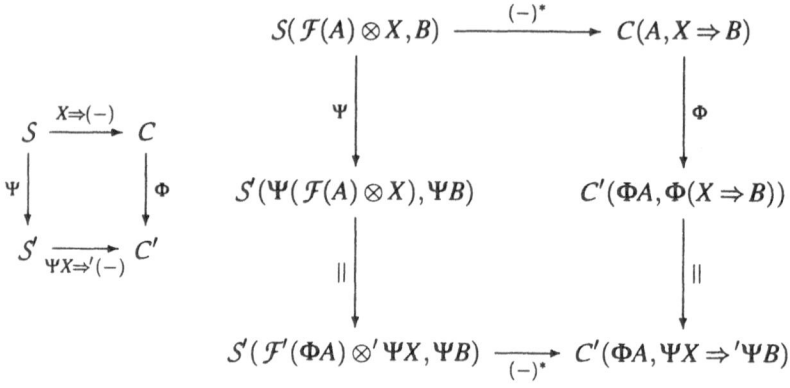

We write **CccSMC** for the 2-category of small cartesian centrally closed symmetric monoidal categories, cartesian centrally closed functors and cartesian-center natural isomorphisms.[1]

As in the last chapter (Lemma 3.2.7 and Proposition 3.2.8) we have a functional completeness result for cartesian centrally closed SMC's:

Lemma 4.2.7 Let $\mathcal{F} : C \to S$ be a cartesian-center SMC and A be an object of C (hence S). Then $\mathcal{F}//A : C//A \to S//A$ is a cartesian centrally-closed SMC. □

Proposition 4.2.8 The obvious identity-on-objects cartesian centrally closed functor $(I, \mathcal{J})$ from $\mathcal{F} : C \to S$ to $\mathcal{F}//A : C//A \to S//A$ satisfies the following universal property. For any identity-on-objects cartesian centrally closed functor (Φ, Ψ) from $\mathcal{F} : C \to S$ to $\mathcal{F}' : C' \to S'$ with an arrow $a : 1 \to A$ in C', there is a unique cartesian centrally closed functor (Φ_a, Ψ_a) from $\mathcal{F}//A : C//A \to S//A$ to $\mathcal{F}' : C' \to S'$ such that $(\Phi_a, \Psi_a) \circ (I, \mathcal{J}) = (\Phi, \Psi)$ and $\Phi_a(\pi_{A,1}) = a$. □

As the first-order case, we introduce the notions of structures and models, and show a soundness property.

Definition 4.2.9 (higher-order sharing structures)

A *higher-order sharing structure* over an S-sorted signature Σ is a sharing structure (Definition 3.2.9) in a strict cartesian centrally closed SMC. □

---

[1] Because of contravariance, for establishing the correspondence with models (Theorem 4.3.3) we need to restrict our attention to isomorphisms; see Chapter 4 of [26] for a related discussion.

Given a higher-order sharing structure $[\![-]\!]$ in a strict cartesian centrally closed SMC $\mathcal{F} : C \to S$, we define $[\![x_1 : \sigma_1, \ldots, x_m : \sigma_m \vdash M : (\tau_1, \ldots, \tau_n)]\!] : [\![\sigma_1]\!] \otimes \ldots \otimes [\![\sigma_m]\!] \to [\![\tau_1]\!] \otimes \ldots \otimes [\![\tau_n]\!]$ in S for each well-typed term $x_1 : \sigma_1, \ldots, x_m : \sigma_m \vdash M : (\tau_1, \ldots, \tau_n)$ as follows, by induction on the typing rules.

$$
\begin{aligned}
[\![\Gamma, x : \sigma \vdash x : (\sigma)]\!] &= \mathcal{F}(\pi') \\
[\![\Gamma \vdash F(M) : (\vec{\tau})]\!] &= [\![\Gamma \vdash M : (\vec{\sigma})]\!] ; [\![F]\!] \\
[\![\Gamma \vdash 0 : ()]\!] &= \mathcal{F}(!) \\
[\![\Gamma \vdash M \otimes N : (\vec{\sigma}, \vec{\tau})]\!] &= \mathcal{F}(\Delta) ; ([\![\Gamma \vdash M : (\vec{\sigma})]\!] \otimes [\![\Gamma \vdash N : (\vec{\tau})]\!]) \\
[\![\Gamma \vdash \text{let } (\vec{x}) \text{ be } M \text{ in } N : (\vec{\tau})]\!] &= \mathcal{F}(\Delta) ; (id \otimes [\![\Gamma \vdash M : (\vec{\sigma})]\!]) ; [\![\Gamma, \vec{x} : \vec{\sigma} \vdash N : (\vec{\tau})]\!] \\
[\![\Gamma \vdash \lambda(\vec{x}).M : ((\vec{\sigma}) \Rightarrow (\vec{\tau}))]\!] &= F([\![\Gamma, \vec{x} : \vec{\sigma} \vdash M : (\vec{\tau})]\!]^*) \\
[\![\Gamma \vdash MN : (\vec{\tau})]\!] &= \mathcal{F}(\Delta) ; ([\![\Gamma \vdash M : ((\vec{\sigma}) \Rightarrow (\vec{\tau}))]\!] \otimes [\![\Gamma \vdash N : (\vec{\sigma})]\!]) ; \mathbf{ap} \\
[\![\Gamma, x' : \sigma', x : \sigma, \Gamma' \vdash M : (\vec{\tau})]\!] &= (id \otimes \mathcal{F}(c) \otimes id) ; [\![\Gamma, x : \sigma, x' : \sigma', \Gamma' \vdash M : (\vec{\tau})]\!]
\end{aligned}
$$

Lemma 4.2.10 Let $[\![-]\!]$ be a higher-order sharing structure. Then $[\![\vec{x} : \vec{\sigma} \vdash M : (\vec{\tau})]\!] = [\![\vec{y} : \vec{\sigma} \vdash M\{\vec{y}/\vec{x}\} : (\vec{\tau})]\!]$ where the x's and y's are disjoint.

Proof: Induction on the construction of M. □

Definition 4.2.11 (higher-order sharing models)
A *higher-order sharing model* of a higher-order acyclic sharing theory is a higher-order sharing structure $[\![-]\!]$ in a strict cartesian centrally closed SMC $\mathcal{F} : C \to S$ such that $[\![\Gamma \vdash M : (\vec{\sigma})]\!] = [\![\Gamma \vdash N : (\vec{\sigma})]\!]$ for each additional $\Gamma \vdash M = N : (\vec{\sigma})$ in the axioms of the theory. □

Thus, again, a higher-order sharing structure automatically induces a model of the pure higher-order acyclic sharing theory.

Theorem 4.2.12 (soundness)
Let $[\![-]\!]$ be a higher-order sharing model of a higher-order sharing theory. If $\Gamma \vdash M = N : (\vec{\sigma})$ is derivable in the theory, then $[\![\Gamma \vdash M : (\vec{\sigma})]\!] = [\![\Gamma \vdash N : (\vec{\sigma})]\!]$.

Proof: We check the additional axioms given in Definition 4.1.2.

- (β):

$$
\begin{aligned}
&[\![\Gamma \vdash (\lambda(\vec{x}).M)N : (\vec{\tau})]\!] \\
= \ &\mathcal{F}\Delta ; (\mathcal{F}([\![\Gamma \vdash \lambda(\vec{x}).M : ((\vec{\sigma}) \Rightarrow (\vec{\tau}))]\!]^*) \otimes [\![\Gamma \vdash N : (\vec{\sigma})]\!]) ; \mathbf{ap} \\
= \ &\mathcal{F}\Delta ; (\mathcal{F}([\![\Gamma, \vec{x} : \vec{\sigma} \vdash M : (\vec{\tau})]\!]^*) \otimes [\![\Gamma \vdash N : (\vec{\sigma})]\!]) ; \mathbf{ap} \\
= \ &\mathcal{F}\Delta ; (id \otimes [\![\Gamma \vdash N : (\vec{\sigma})]\!]) ; [\![\Gamma, \vec{x} : \vec{\sigma} \vdash M : (\vec{\tau})]\!] \\
= \ &[\![\Gamma \vdash \text{let } (\vec{x}) \text{ be } N \text{ in } M : (\vec{\tau})]\!]
\end{aligned}
$$

- (η_0):

$$
\begin{aligned}
&[\![\Gamma \vdash \lambda(\vec{x}).y\vec{x} : ((\vec{\sigma}) \Rightarrow (\vec{\tau}))]\!] \\
= \ &\mathcal{F}([\![\Gamma, \vec{x} : \vec{\sigma} \vdash y\vec{x} : (\vec{\tau})]\!]^*) \\
= \ &\mathcal{F}((\mathcal{F}\Delta ; ([\![\Gamma, \vec{x} : \vec{\sigma} \vdash y : ((\vec{\sigma}) \Rightarrow (\vec{\tau}))]\!] \otimes [\![\Gamma, \vec{x} : \vec{\sigma} \vdash \vec{x} : (\vec{\sigma})]\!]) ; \mathbf{ap})^*) \\
= \ &\mathcal{F}((([\![\Gamma \vdash y : ((\vec{\sigma}) \Rightarrow (\vec{\tau}))]\!] \otimes [\![\vec{x} : \vec{\sigma} \vdash \vec{x} : (\vec{\sigma})]\!]) ; \mathbf{ap})^*) \\
= \ &[\![\Gamma \vdash y : ((\vec{\sigma}) \Rightarrow (\vec{\tau}))]\!] ; \mathcal{F}(\mathbf{ap}^*) \\
= \ &[\![\Gamma \vdash y : ((\vec{\sigma}) \Rightarrow (\vec{\tau}))]\!]
\end{aligned}
$$

- (σ_v):

$$
\begin{aligned}
& [\![\Gamma \vdash \text{let } (x) \text{ be } \lambda(\vec{y}).M \text{ in } N : (\vec{\tau})]\!] \\
= \; & \mathcal{F}\Delta; (id \otimes [\![\Gamma \vdash \lambda(\vec{y}).M : ((\vec{\sigma}) \Rightarrow (\vec{\sigma}'))]\!]); [\![\Gamma, x : (\vec{\sigma}) \Rightarrow (\vec{\sigma}') \vdash N : (\vec{\tau})]\!] \\
= \; & \mathcal{F}\Delta; (id \otimes \mathcal{F}([\![\Gamma, \vec{y} : \vec{\sigma} \vdash M : (\vec{\sigma}')]\!]^*)); [\![\Gamma, x : (\vec{\sigma}) \Rightarrow (\vec{\sigma}') \vdash N : (\vec{\tau})]\!]
\end{aligned}
$$

The rest is similar to the case of (σ_{var}), by induction on the construction of N.

- (app_1):

$$
\begin{aligned}
& [\![\Gamma \vdash (\text{let } (\vec{x}) \text{ be } L \text{ in } M)N : (\vec{\tau})]\!] \\
= \; & \mathcal{F}\Delta; ([\![\Gamma \vdash \text{let } (\vec{x}) \text{ be } L \text{ in } M : ((\vec{\sigma}) \Rightarrow (\vec{\tau}))]\!] \otimes [\![\Gamma \vdash N : (\vec{\sigma})]\!]); \mathbf{ap} \\
= \; & \mathcal{F}\Delta; ((\mathcal{F}\Delta; (id \otimes \Gamma \vdash L : (\vec{\sigma}')); [\![\Gamma, \vec{x} : \vec{\sigma}' \vdash M : ((\vec{\sigma}) \Rightarrow (\vec{\tau}))]\!]) \\
& \otimes [\![\Gamma \vdash N : (\vec{\sigma})]\!]); \mathbf{ap} \\
= \; & \mathcal{F}\Delta; (id \otimes \Gamma \vdash L : (\vec{\sigma}')); \mathcal{F}\Delta; \\
& ([\![\Gamma, \vec{x} : \vec{\sigma}' \vdash M : ((\vec{\sigma}) \Rightarrow (\vec{\tau}))]\!] \otimes [\![\Gamma, \vec{x} : \vec{\sigma}' \vdash N : (\vec{\sigma})]\!]); \mathbf{ap} \\
= \; & \mathcal{F}\Delta; (id \otimes \Gamma \vdash L : (\vec{\sigma}')); [\![\Gamma, \vec{x} : \vec{\sigma}' \vdash MN : (\vec{\tau})]\!] \\
= \; & [\![\Gamma \vdash \text{let } (\vec{x}) \text{ be } L \text{ in } MN : (\vec{\tau})]\!]
\end{aligned}
$$

- (app_2) is similar to (app_1).

$\square$

For a higher-order sharing theory $\mathbb{T}$ (determined by a set of axioms) over a signature Σ, we write **HSharingMod**$(\mathbb{T}, (\mathcal{F} : C \to S))$ for the category of $\mathbb{T}$'s higher-order sharing models in a strict cartesian centrally closed SMC $\mathcal{F} : C \to S$ and the homomorphisms whose components are isomorphisms (see the footnote in page 60).

4.3 The Classifying Category

As in the first-order case, we construct a term model from a higher-order sharing theory.

Proposition 4.3.1 Given a higher-order acyclic sharing theory $\mathbb{T}$, we construct a strict cartesian centrally closed SMC $\mathcal{F}_{\mathbb{T}} : C_{\mathbb{T}} \to S_{\mathbb{T}}$ as follows. $S_{\mathbb{T}}$ is given in the same way as the first-order case, thus arrows are equivalence classes of well-typed terms with contexts. $C_{\mathbb{T}}$ is the subcategory of $S_{\mathbb{T}}$ whose arrows are equivalence classes of well-typed values (Definition 4.1.3) with contexts. $\mathcal{F}_{\mathbb{T}}$ is then the inclusion functor. $\square$

Proof: To see that $C_{\mathbb{T}}$ is a (strict) cartesian category is straightforward. The essential point is the verification of centrally closedness. For each object $(\vec{\sigma})$ define $(\vec{\sigma}) \Rightarrow (-) : S_{\mathbb{T}} \to C_{\mathbb{T}}$ by

$$(\vec{\sigma}) \Rightarrow [\vec{x} : \vec{\sigma}' \vdash M : (\vec{\tau})] = [f : (\vec{\sigma}) \Rightarrow (\vec{\sigma}') \vdash \lambda(\vec{y}).(\text{let } (\vec{x}) \text{ be } f(\vec{y}) \text{ in } M) : ((\vec{\sigma}) \Rightarrow (\vec{\tau}))].$$

We need to check that $(\vec{\sigma}) \Rightarrow (-)$ is indeed a functor:

$$
\begin{aligned}
& (\vec{\sigma}) \Rightarrow [\vec{x} : \vec{\tau} \vdash \vec{x} : (\vec{\tau})] \\
= \; & [f : (\vec{\sigma}) \Rightarrow (\vec{\tau}) \vdash \lambda(\vec{y}).(\text{let } (\vec{x}) \text{ be } f(\vec{y}) \text{ in } \vec{x}) : ((\vec{\sigma}) \Rightarrow (\vec{\tau}))] \\
= \; & [f : (\vec{\sigma}) \Rightarrow (\vec{\tau}) \vdash \lambda(\vec{y}).f(\vec{y}) : ((\vec{\sigma}) \Rightarrow (\vec{\tau}))] && (\text{id}) \\
= \; & [f : (\vec{\sigma}) \Rightarrow (\vec{\tau}) \vdash f : ((\vec{\sigma}) \Rightarrow (\vec{\tau}))] && (\eta_0)
\end{aligned}
$$

$$((\vec{\sigma}) \Rightarrow [\vec{x} : \vec{\tau} \vdash M : (\vec{\tau}')]) ; ((\vec{\sigma}) \Rightarrow [\vec{y} : \vec{\tau}' \vdash N : (\vec{\tau}'')])$$
$$= \quad [f : (\vec{\sigma}) \Rightarrow (\vec{\tau}) \vdash \lambda(\vec{u}).(\text{let } (\vec{x}) \text{ be } f(\vec{u}) \text{ in } M) : ((\vec{\sigma}) \Rightarrow (\vec{\tau}'))] ;$$
$$[g : (\vec{\sigma}) \Rightarrow (\vec{\tau}') \vdash \lambda(\vec{v}).(\text{let } (\vec{y}) \text{ be } g(\vec{v}) \text{ in } N) : ((\vec{\sigma}) \Rightarrow (\vec{\tau}''))]$$
$$= \quad [f : (\vec{\sigma}) \Rightarrow (\vec{\tau}) \vdash$$
$$\text{let } g \text{ be } \lambda(\vec{u}).(\text{let } (\vec{x}) \text{ be } f(\vec{u}) \text{ in } M) \text{ in } \lambda(\vec{v}).(\text{let } (\vec{y}) \text{ be } g(\vec{v}) \text{ in } N)$$
$$: ((\vec{\sigma}) \Rightarrow (\vec{\tau}''))]$$
$$= \quad [f : (\vec{\sigma}) \Rightarrow (\vec{\tau}) \vdash$$
$$\lambda(\vec{v}).(\text{let } (\vec{y}) \text{ be } (\lambda(\vec{u}).(\text{let } (\vec{x}) \text{ be } f(\vec{u}) \text{ in } M))(\vec{v}) \text{ in } N)$$
$$: ((\vec{\sigma}) \Rightarrow (\vec{\tau}''))] \qquad\qquad (\sigma_v)$$
$$= \quad [f : (\vec{\sigma}) \Rightarrow (\vec{\tau}) \vdash$$
$$\lambda(\vec{v}).(\text{let } (\vec{y}) \text{ be } (\text{let } (\vec{x}) \text{ be } f(\vec{v}) \text{ in } M) \text{ in } N) : ((\vec{\sigma}) \Rightarrow (\vec{\tau}''))] \qquad (\beta_v)$$
$$= \quad [f : (\vec{\sigma}) \Rightarrow (\vec{\tau}) \vdash$$
$$\lambda(\vec{v}).(\text{let } (\vec{x}) \text{ be } f(\vec{v}) \text{ in let } (\vec{y}) \text{ be } M \text{ in } N) : ((\vec{\sigma}) \Rightarrow (\vec{\tau}''))] \qquad (\text{ass}_1)$$
$$= \quad (\vec{\sigma}) \Rightarrow [\vec{x} : \vec{\sigma} \vdash \text{let } (\vec{y}) \text{ be } M \text{ in } N : (\vec{\tau}')]$$
$$= \quad (\vec{\sigma}) \Rightarrow ([\vec{x} : \vec{\tau} \vdash M : (\vec{\tau}')] ; [\vec{y} : \vec{\tau}' \vdash N : (\vec{\tau}'')])$$

Now we check that $(\vec{\sigma}) \Rightarrow (-)$ is a right adjoint of $\mathcal{F}_{\mathbb{T}}(-) \otimes (\vec{\sigma})$. Given $f = [\Gamma, \vec{x} : \vec{\sigma} \vdash M : \vec{\tau}]$, define $f^* = [\Gamma \vdash \lambda(\vec{x}).M : ((\vec{\sigma}) \Rightarrow (\vec{\tau}))]$. Also define $\mathbf{ap} = [[f : ((\vec{\sigma}) \Rightarrow (\vec{\tau})), \vec{x} : \vec{\sigma} \vdash f(\vec{x}) : (\vec{\tau})]]$. Then $(\mathcal{F}_{\mathbb{T}}(f^*) \otimes id) ; \mathbf{ap} = f$ as

$$\quad LHS$$
$$= \quad [\Gamma, \vec{y} : \vec{\sigma} \vdash \text{let } (f, \vec{z}) \text{ be } (\lambda(\vec{x}).M) \otimes \vec{y} \text{ in } f(\vec{z}) : (\vec{\tau})]$$
$$= \quad [\Gamma, \vec{y} : \vec{\sigma} \vdash \text{let } (f) \text{ be } \lambda(\vec{x}).M \text{ in let } (\vec{z}) \text{ be } \vec{y} \text{ in } f(\vec{z}) : (\vec{\tau})]$$
$$= \quad [\Gamma, \vec{y} : \vec{\sigma} \vdash (\lambda(\vec{x}).M)(\vec{y}) : (\vec{\tau})] \qquad\qquad (\sigma_{\text{var}}), (\sigma_v)$$
$$= \quad [\Gamma, \vec{y} : \vec{\sigma} \vdash M\{\vec{y}/\vec{x}\} : (\vec{\tau})] \qquad\qquad (\beta_v)$$
$$= \quad RHS$$

Also, for $g = [\Gamma \vdash V : ((\vec{\sigma}) \Rightarrow (\vec{\tau}))]$ in $C_{\mathbb{T}}$ (thus V is equal to either a variable or a lambda abstraction), we show that $((\mathcal{F}_{\mathbb{T}}(g) \otimes id_{(\vec{\sigma})}) ; \mathbf{ap})^* = g$, which completes the proof of adjointness.

$$\quad LHS$$
$$= \quad [\Gamma, \vec{x} : \vec{\sigma} \vdash \text{let } (f, \vec{y}) \text{ be } V \otimes \vec{x} \text{ in } f(\vec{y}) : (\vec{\tau})]^*$$
$$= \quad [\Gamma, \vec{x} : \vec{\sigma} \vdash V(\vec{x}) : (\vec{\tau})]^* \qquad\qquad (\text{ass}_2), (\sigma_v) \text{ and } (\sigma_{\text{var}})$$
$$= \quad [\Gamma \vdash \lambda(\vec{x}).V(\vec{x}) : ((\vec{\sigma}) \Rightarrow (\vec{\tau}))]$$
$$= \quad RHS \qquad\qquad (\eta_v)$$

$\square$

Now we have parallel results to the first-order case:

Theorem 4.3.2 (completeness)
Given a higher-order acyclic sharing theory $\mathbb{T}$, there is a complete model in $\mathcal{F}_{\mathbb{T}} : C_{\mathbb{T}} \to S_{\mathbb{T}}$, given by $[\![\sigma]\!] = \sigma$ and $[\![F]\!] = [\vec{x} : \vec{\sigma} \vdash F(\vec{x}) : (\vec{\tau})]$ for $F : (\vec{\sigma}) \to (\vec{\tau})$. $\square$

Theorem 4.3.3

$$\mathbf{CccSMC}((C_{\mathbb{T}} \overset{\mathcal{F}_{\mathbb{T}}}{\to} S_{\mathbb{T}}), (C \overset{\mathcal{F}}{\to} S)) \simeq \mathbf{HSharingMod}(\mathbb{T}, (C \overset{\mathcal{F}}{\to} S)).$$

$\square$

5
Relating Models

One advantage of our axiomatic approach to the semantic models of sharing theories is that it enables us to compare similar systems arising from computer science by relating the classes of models. In this chapter we give a few case studies: we relate the first-order acyclic sharing theories and the higher-order ones; and our sharing theories with those of Moggi's *notions of computation* [71, 72]; and also with *intuitionistic linear type theory* [12, 18, 20].

All the techniques involved here are fairly standard (though we review all the needed notions below), and we believe that our semantic comparisons give a clear account of often too complicated syntactic translations and considerations.

The content of this chapter is relatively independent of the later chapters. The semantic proofs of the conservativity results are also reported in the joint work with Barber, Gardner and Plotkin [13, 35] for the corresponding action calculi (see Chapter 8). The detailed descriptions of syntactic translations are found in these papers.

5.1 Preliminaries from Category Theory

We supply additional material from category theory which is needed in this chapter.

Definition 5.1.1 (monoidal adjunction)
A *(symmetric) monoidal adjunction* between (symmetric) monoidal categories is an adjunction in which both of the functors are (symmetric) monoidal and the unit and counit are monoidal natural transformations. □

Proposition 5.1.2 (Kelly [53])
The left adjoint part of a (symmetric) monoidal adjunction is necessarily strong (symmetric) monoidal. Conversely, if a strong (symmetric) monoidal functor has a right adjoint, then the adjunction is (symmetric) monoidal. □

Definition 5.1.3 (strong monad [55])
A *strong monad* over a symmetric monoidal category $\mathcal{M} = (\mathcal{M}, \otimes, I, a, l, r, c)$ is a monad (T, η, μ) on $\mathcal{M}$ with a natural transformation (called a *tensorial strength*)

$$\theta_{A,X} : A \otimes TX \to T(A \otimes X)$$

subject to the following commutative diagrams.

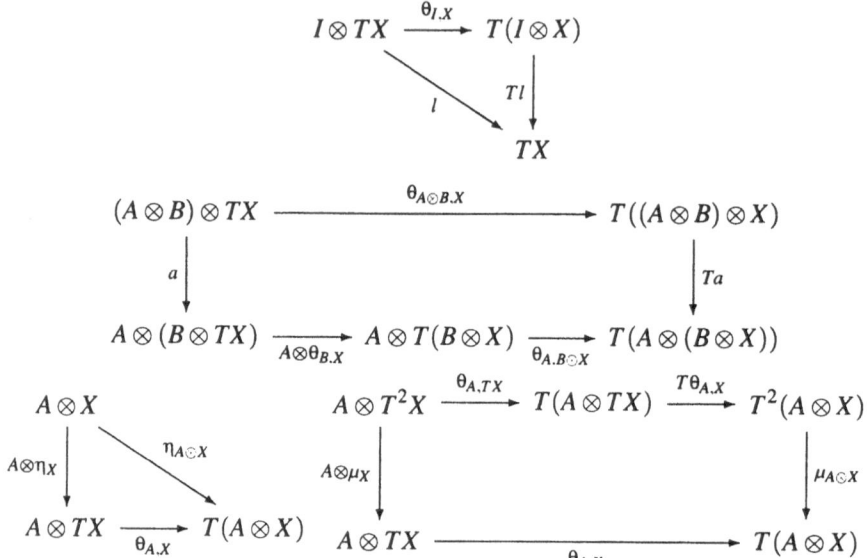

If the following diagram commutes, T is said to be *commutative* (and θ is a *commutative strength*):

$$
\begin{array}{ccccccc}
TA \otimes TB & \xrightarrow{\theta_{TA,B}} & T(TA \otimes B) & \xrightarrow{Tc} & T(B \otimes TA) & \xrightarrow{T\theta_{B,A}} & T^2(B \otimes A) \\
\downarrow{\scriptstyle c} & & & & & & \downarrow{\scriptstyle T^2 c} \\
TB \otimes TA & & & & & & T^2(A \otimes B) \\
\downarrow{\scriptstyle \theta_{TB,A}} & & & & & & \downarrow{\scriptstyle \mu} \\
T(TB \otimes A) & \xrightarrow{Tc} & T(A \otimes TB) & \xrightarrow{T\theta_{A,B}} & T^2(A \otimes B) & \xrightarrow{\mu} & T(A \otimes B)
\end{array}
$$

$\square$

5.2 Higher-Order Extension

Since the pure higher-order sharing theory is obtained from the first-order one by adding additional term constructs and axioms, there is an obvious sound translation from the first-order theory to the higher-order theory. We show that this translation is not only sound but also conservative (faithful) and full. It is possible to prove this by a purely syntactic way (by comparing the normal forms), but here we give a simple semantic proof using a model embedding technique.

We start with a general fact on the Yoneda construction (free cocompletion) on symmetric monoidal categories [29]. A systematic account can be found in [77].

Lemma 5.2.1 Let C, $\mathcal{D}$ be small symmetric monoidal categories, with a strict symmetric monoidal functor $\mathcal{F}: C \to \mathcal{D}$ which is identity on objects. Then there exist small symmetric monoidal categories $\bar{C}$ and $\bar{\mathcal{D}}$, an identity-on-objects strict symmetric monoidal functor $\bar{\mathcal{F}}: \bar{C} \to \bar{\mathcal{D}}$, together with fully faithful strict symmetric monoidal functors $i_C: C \to \bar{C}$ and $i_{\mathcal{D}}: \mathcal{D} \to \bar{\mathcal{D}}$ such that the induced square commutes and $\bar{\mathcal{F}}$ has a right adjoint; moreover $\bar{C}$ is symmetric monoidal closed, and i_C is dense.

Proof: Let $\bar{C}$ be the presheaf category $[C^{\mathrm{op}}, \mathbf{Set}]$ and in_C be the Yoneda embedding. As is well-known, $\bar{C}$ is the free symmetric monoidal cocompletion of C and in_C is strict symmetric monoidal [29, 44]. Then F extends to a strict symmetric monoidal functor $\hat{F}: \bar{C} \longrightarrow [\mathcal{D}^{\mathrm{op}}, \mathbf{Set}]$ with a right adjoint $U = [F^{\mathrm{op}}, \mathbf{Set}]$, so that $\hat{F}$ strictly commutes with F. Although $\hat{F}$ may not be identity-on-objects, we can factorize it as $\hat{F} = J \circ \bar{F}$ so that $\bar{F}: \bar{C} \longrightarrow \bar{\mathcal{D}}$ is identity-on-objects and $J: \bar{\mathcal{D}} \longrightarrow [\mathcal{D}^{\mathrm{op}}, \mathbf{Set}]$ is fully faithful. A right adjoint of $\bar{F}$ is then given by $U \circ J: \bar{\mathcal{D}} \longrightarrow \bar{C}$. ($\bar{C}$ and $\bar{\mathcal{D}}$ obtained are not small, but we can cut them down to be closed and small.) ◻

Corollary 5.2.2 Let $\mathcal{F}: C \to S$ be a cartesian-center SMC. Then there is a cartesian centrally closed SMC $\bar{\mathcal{F}}: \bar{C} \to \bar{S}$ with a fully faithful cartesian-center functor (i_C, i_S) from $\mathcal{F}: C \to S$ to $\bar{\mathcal{F}}: \bar{C} \to \bar{S}$. Moreover i_C is dense. ◻

Theorem 5.2.3 (conservativity)
If $\Gamma \vdash M = N: (\vec{\sigma})$ is derivable in the pure higher-order acyclic sharing theory, then it is also derivable in the pure acyclic sharing theory.

Proof: Let $\mathcal{F}_{\mathrm{T}}: C_{\mathrm{T}} \to S_{\mathrm{T}}$ be the classifying category of the pure acyclic sharing theory. From this, we get a cartesian centrally closed SMC $\bar{\mathcal{F}}_{\mathrm{T}}: \bar{C}_{\mathrm{T}} \to \bar{S}_{\mathrm{T}}$ to which the classifying category fully and faithfully embeds. (This is not strict, so we need to take its strict equivalent to be more precise.) The sharing structure in $\mathcal{F}_{\mathrm{T}}: C_{\mathrm{T}} \to S_{\mathrm{T}}$ canonically extends to a higher-order sharing structure in $\bar{\mathcal{F}}_{\mathrm{T}}: \bar{C}_{\mathrm{T}} \to \bar{S}_{\mathrm{T}}$, and it is routine to see that it induces a higher-order sharing model. Then the fully faithful embedding from $\mathcal{F}_{\mathrm{T}}: C_{\mathrm{T}} \to S_{\mathrm{T}}$ to $\bar{\mathcal{F}}_{\mathrm{T}}: \bar{C}_{\mathrm{T}} \to \bar{S}_{\mathrm{T}}$ factors through the classifying category of the pure higher-order acyclic sharing theory, and this implies that the canonical translation from $\mathcal{F}_{\mathrm{T}}: C_{\mathrm{T}} \to S_{\mathrm{T}}$ to the classifying category of the higher-order theory is faithful. ◻

5.3 Notions of Computation

It has been pointed that Moggi's computational lambda calculus [71] looks like higher-order graph rewriting systems. And we notice that our higher-order acyclic sharing theory is fairly similar to Moggi's calculus. We make this intuition precise by observing that Moggi's models and ours are essentially the same. This comparison can be understood as an instance of the work by Power and Robinson [77], where they give a reformulation of Moggi's notions of computation in terms of premonoidal categories. In this thesis we use just symmetric monoidal categories, but the story presented here is inspired from their general framework.

Definition 5.3.1 (λ_c-models [71])
Let C be a cartesian category. A λ_c-*model over* C is a strong monad (T,η,μ,θ) which satisfies the mono-requirement (each component of η is mono) and has Kleisli exponents, i.e. there is a chosen object $A \Rightarrow B$ for each object A and B and a natural isomorphism

$$C_T(J(- \times A), -) \simeq C(-, A \Rightarrow -)$$

where C_T is the Kleisli category of T and $J : C \longrightarrow C_T$ is given by $J(f) = f;\eta$. $\qquad\square$

Definition 5.3.2 A λ_c-model is said to be *commutative* if the tensorial strength is commutative. $\qquad\square$

Theorem 5.3.3 There is a bijective correspondence between commutative λ_c-models and faithful cartesian centrally closed SMC's.

<u>Proof:</u> Routine; see for instance [77]. $\qquad\square$

We reproduce Moggi's computational lambda calculus (λ_c-calculus) [71] (the simply typed version in [60]) as below.

Definition 5.3.4 (computational lambda calculus)
The *computational lambda calculus* (λ_c-*calculus*) [71, 60] is determined by the following data.
[Types]

$$\sigma,\tau\ldots \;::=\; b \mid \sigma \Rightarrow \tau \quad \text{(where } b \text{ is a base type)}$$

[Syntactic domains]

Variables	$x, y, z \ldots$
Raw Terms	$M, N \ldots \;::=\; x \mid \lambda x.M \mid MN \mid \text{let } x = M \text{ in } N$
Values	$V, W \ldots \;::=\; x \mid \lambda x.M$

[Typing]

$$\frac{}{\Gamma, x : \sigma \vdash x : \sigma} \qquad \frac{\Gamma, x : \sigma \vdash M : \tau}{\Gamma \vdash \lambda x.M : \sigma \Rightarrow \tau}$$

$$\frac{\Gamma \vdash M : \sigma \Rightarrow \tau \quad \Gamma \vdash N : \sigma}{\Gamma \vdash MN : \tau} \qquad \frac{\Gamma \vdash M : \sigma \quad \Gamma, x : \sigma \vdash N : \tau}{\Gamma \vdash \text{let } x = M \text{ in } N : \tau}$$

[Axioms]

(β_v)	$(\lambda x.M)V$	$=$	$M\{V/x\}$
(η_v)	$\lambda x.(Vx)$	$=$	V
			$(x \notin FV(V))$
(let$_v$)	$\text{let } x = V \text{ in } M$	$=$	$M\{V/x\}$
(id)	$\text{let } x = M \text{ in } x$	$=$	M
(comp)	$\text{let } y = (\text{let } x = L \text{ in } M) \text{ in } N$	$=$	$\text{let } x = L \text{ in let } y = M \text{ in } N$
(let.1)	EM	$=$	$\text{let } z = E \text{ in } zM$
(let.2)	VE	$=$	$\text{let } x = E \text{ in } Vx$

where E ranges over non-values, i.e. applications and let-blocks. ☐

Definition 5.3.5 The *commutative λ_c-calculus* is obtained by assuming the following additional axiom.

(comm)   let $x = L$ in let $y = M$ in N $=$ let $y = M$ in let $x = L$ in N

☐

Since Moggi's (commutative) calculus is sound and complete for (commutative) λ_c-models, we have

Theorem 5.3.6 The pure higher-order acyclic sharing theory is a conservative extension of the commutative computational lambda calculus. ☐

5.4   Models of Intuitionistic Linear Logic

A model of propositional intuitionistic linear logic may be described as a symmetric monoidal adjunction between a cartesian closed category and a symmetric monoidal closed category [12, 19, 18, 20]. Such a structure induces a cartesian-center SMC, i.e. a model of the acyclic sharing theory, which implies that there is a sound interpretation from the sharing theory into the term calculus of intuitionistic type theory. We show that it is conservative, thus a linear type theory is seen as a conservative extension of the pure acyclic sharing theory. Again we avoid syntactic proof, by relating the classes of models.

Definition 5.4.1 (LNL models [18])

A *linear/non-linear model* (*LNL model*) is a symmetric monoidal adjunction $C \underset{G}{\overset{F}{\underset{\perp}{\rightleftarrows}}} S$ where C is a cartesian closed category and S is a symmetric monoidal closed category. ☐

First, it is fairly easy to see that an LNL model induces a cartesian centrally closed SMC (equivalently Moggi's model, as observed in [19]).

Lemma 5.4.2 Let $C \underset{G}{\overset{F}{\underset{\perp}{\rightleftarrows}}} S$ be an LNL model. Consider the monad $T = G \circ F$ and its Kleisli category C_T. Then the canonical functor $J : C \to C_T$ is a cartesian centrally closed SMC; and C_T is a full subcategory of S; and F factors through J. ☐

Intuitively, C_T corresponds to a fragment of the linear type theory in which types are of the form $!A$. Therefore a (higher-order) sharing theory can be translated into this fragment of a linear type theory (equipped with constants corresponding to operator symbols). See [18, 19] for details of the term calculus.

The conservativity of the translation is a consequence of the following construction.

Theorem 5.4.3 Given a cartesian-center SMC $\mathcal{F} : C \to S$, there is an LNL model $\widehat{C} \xrightarrow[\underset{U}{\longleftarrow}]{\widehat{\mathcal{F}}} \widehat{S}$ such that there are fully faithful strict symmetric monoidal functors $j : C \to \widehat{C}$ and $j' : S \to \widehat{S}$ which satisfy $j' \circ \mathcal{F} = \widehat{\mathcal{F}} \circ j$.

Proof: Similar to Lemma 5.2.1.   □

Theorem 5.4.4 The interpretation from the pure acyclic sharing theory into the LNL logic (with operator symbols of Σ) is conservative.   □

6
Models of Cyclic Sharing Theory

We have delayed introducing the models for cyclic sharing graphs until now. The main reason is that, while the models of acyclic settings are obtained by revising well known concepts from category theory, we need a relatively new notion for interpreting cyclic bindings - *traced monoidal categories* introduced by Joyal, Street and Verity [50]. The notion of trace, while the concept itself goes back to the classical traces of linear maps between finite dimensional vector spaces, has originally been invented for analyzing cyclic structures arising from mathematics and physics, notably the interplay of low-dimensional topology (knot theory) and quantum groups (e.g. [78, 51]); it is then a natural idea to use this concept for modeling our cyclic graph structure too, and it does work.

In this chapter we deal with the models of cyclic sharing theories. The additional structure we require for modeling cyclic bindings is trace as mentioned above, which we review below. The construction of this chapter is therefore parallel to that for models of acyclic sharing theories (Chapter 3), with additional considerations on the trace structure induced by cyclic bindings.

6.1 Traced Monoidal Categories

The notion of trace we give here for symmetric monoidal categories is adopted from the original definition of trace for balanced monoidal categories [49] in [50].

For ease of presentation, in this section we write as if our monoidal categories are strict (i.e. monoidal products are strictly associative and coherence isomorphisms are identities).

Definition 6.1.1 (Traced symmetric monoidal categories [50])
A symmetric monoidal category $(C, \otimes, I, c)$ (where c is the symmetry; $c_{X,Y} : X \otimes Y \longrightarrow Y \otimes X$) is said to be *traced* if it is equipped with a natural family of functions, called a *trace*,

$$Tr_{A,B}^X : C(A \otimes X, B \otimes X) \longrightarrow C(A, B)$$

subject to the following three conditions.

- **Vanishing:**
$$Tr_{A,B}^I(f) = f : A \to B$$

  where $f : A \longrightarrow B$, and

$$Tr_{A,B}^{X \otimes Y}(f) = Tr_{A,B}^X(Tr_{A \otimes X, B \otimes X}^Y(f)) : A \longrightarrow B$$

  where $f : A \otimes X \otimes Y \longrightarrow B \otimes X \otimes Y$

- **Superposing:**

$$Tr^X_{C\otimes A,C\otimes B}(id_C \otimes f) = id_C \otimes Tr^X_{A,B}(f) : C\otimes A \longrightarrow C\otimes B$$

where $f : A\otimes X \longrightarrow B\otimes X$

- **Yanking:**

$$Tr^X_{X,X}(c_{X,X}) = id_X : X \longrightarrow X \qquad \square$$

We may omit the subscripts if there is no confusion.

We present the graphical version of these axioms to help with the intuition of traced categories as categories with cycles (or feedback, reflexion). Such graphical languages for various monoidal categories have been developed in [48].

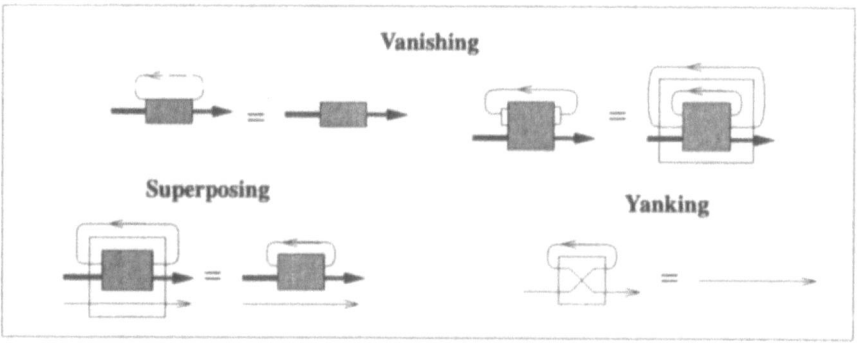

Note that naturality of a trace can be axiomatized as follows.

- Naturality in A **(Left Tightening)**

$$Tr^X_{A,B}((g\otimes id_X);f) = g;Tr^X_{A',B}(f) : A \longrightarrow B$$

where $f : A'\otimes X \longrightarrow B\otimes X, g : A \longrightarrow A'$

- Naturality in B **(Right Tightening)**

$$Tr^X_{A,B}(f;(g\otimes id_X)) = Tr^X_{A,B'}(f);g : A \longrightarrow B$$

where $f : A\otimes X \longrightarrow B'\otimes X, g : B' \longrightarrow B$

- Naturality in X **(Sliding)**

$$Tr^X_{A,B}(f;(id_B\otimes g)) = Tr^{X'}_{A,B}((id_A\otimes g);f) : A \longrightarrow B$$

where $f : A\otimes X \longrightarrow B\otimes X', g : X' \longrightarrow X$

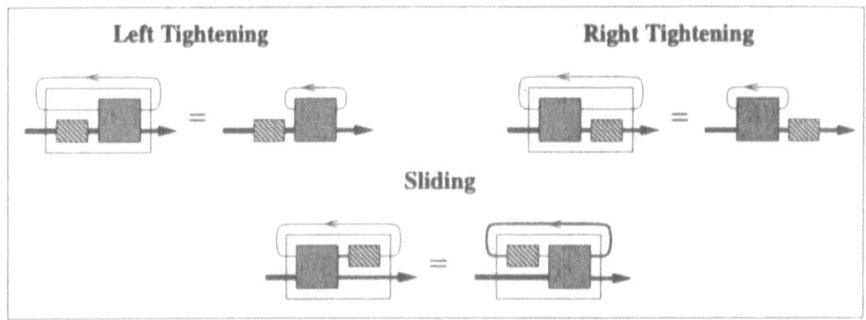

Remark 6.1.2 The axiom **Superposing** is slightly simplified from the original version in [50]

$$Tr^X_{A\otimes C,B\otimes D}((id_A \otimes c_{C,X});(f\otimes g);(id_B \otimes c_{X,D})) = Tr^X_{A,B}(f)\otimes g$$

where $f : A\otimes X \longrightarrow B\otimes X, g : C \longrightarrow D$. Assuming axioms **Left & Right Tightenings**, ours is derivable from this original one, and vice versa. □

Example 6.1.3 A *compact closed category* [52, 54] is a symmetric monoidal category with a contravariant endofunctor $(-)^*$ and natural transformations $\eta_A : I \to A\otimes A^*$ and $\varepsilon_A : A^* \otimes A \to I$ such that

$$A \xrightarrow{r^{-1}_A} I\otimes A \xrightarrow{\eta_A \otimes A} (A\otimes A^*)\otimes A \xrightarrow{a_{A,A^*,A}} A\otimes (A^*\otimes A) \xrightarrow{A\otimes \varepsilon_A} A\otimes I \xrightarrow{l_A} A$$

$$A^* \xrightarrow{l^{-1}_{A^*}} A^*\otimes I \xrightarrow{A^*\otimes \eta_A} A^*\otimes (A\otimes A^*) \xrightarrow{a^{-1}_{A^*,A,A^*}} (A^*\otimes A)\otimes A^* \xrightarrow{\varepsilon_A \otimes A^*} I\otimes A^* \xrightarrow{r_{A^*}} A^*$$

agree with identity arrows. Such a category is closed, with exponents $[A,B]$ given by $A^* \otimes B$. In [50], it is shown that any compact closed category is traced, for instance the category of sets and binary relations, and the category of finite dimensional vector spaces (see examples in the next chapter). Moreover, the structure theorem (which gives a more general relation between traced balanced monoidal categories and tortile monoidal categories [80]) in [50] tells us that any traced symmetric monoidal category can be fully and faithfully embedded into a compact closed category (which can be obtained by a simple fraction construction). This fact, however, does not imply that the usage of traced categories is the same as that of compact closed categories. For the study of cyclic data structures, we find traced categories more useful than compact closed categories, as the latter seems to be too strong for our purpose to model cyclic structures (rather than dualities which we do not need at least a priori). In particular, while we will see that there are many interesting traced cartesian categories, a compact closed category whose monoidal product is cartesian is trivial (because, for such C, we have $C(A,B) \simeq C(A\otimes B^*, 1)$). □

Example 6.1.4 Examples which are not compact closed include the category of sets and partial functions with coproduct as monoidal product, and the category of sets and binary relations with biproduct as monoidal product (described in detail in [50]). More examples will be introduced in this and next chapters. □

We also refer Abramsky's survey [2] for some computer science oriented examples, especially some models related to Girard's Geometry of Interaction [37].

Let us briefly answer a few frequently-asked-questions on the axiomatization of traces, as this might be helpful for avoiding some misunderstanding.

Fact. $Tr^X id_X = id_I$ is not always true.   □

This condition is true for all traced cartesian categories and also for many other traced categories, for instance the category of sets and binary relations, but we also have lots of counterexamples - a basic one is the category of finite dimensional vector spaces and linear maps, which is the prototypical traced monoidal category (traced in the very classical sense), where $Tr^X id_X$ is the function which multiplies the dimension of X, and it is therefore not the identity id_I unless X's dimension is 1.

A reflexive action calculus (to be introduced in Chapter 8) does not satisfy this condition. However, as suggested and discussed in [67, 61], assuming this condition may add some desirable strength to the calculus, since an action (process) of the form $Tr^X id_X$ seems to have no computational significance and hence can be identified with the "empty action" id_I.

Fact. A trace may not be functorial – that is, $Tr^X(f;g) = Tr^X f; Tr^X g$ does not hold in general.   □

Here is evidence that an interesting trace is unlikely to be functorial:

Proposition 6.1.5 For any traced symmetric monoidal category, the functoriality condition

$$Tr^X(f;g) = Tr^X f; Tr^X g \text{ for any } X, f, g$$

is true if and only if

$$c_{X,X} = id_{X \otimes X} \text{ for any } X.$$

<u>Proof:</u> See Appendix.   □

That is, if a symmetric monoidal category has a functorial trace, the symmetries of the form $c_{X,X}$ must be the identities – not many nontrivial examples seem to satisfy such a condition.

Finally we shall mention an observation by Plotkin:

Fact. Traced symmetric monoidal categories are "cancellable", in the sense that if $f \otimes id_{X \otimes X} = g \otimes id_{X \otimes X}$ then $f \otimes id_X = g \otimes id_X$.   □

This immediately follows from $f \otimes id_X = Tr^X((f \otimes id_{X \otimes X}); (id \otimes c_{X,X}))$. As there are symmetric monoidal categories which are not cancellable, we have

Theorem 6.1.6 (Plotkin)
There are symmetric monoidal categories to which we cannot add traces freely without

causing a collapse. In other words, there are symmetric monoidal categories which cannot be faithfully embedded into traced monoidal categories by strong symmetric monoidal functors. □

Plotkin conjectures that cancellable symmetric monoidal categories (which include all cartesian categories as well as all cartesian-center SMC's) faithfully embed into traced categories.

Definition 6.1.7 (traced functors [50])
A strong symmetric monoidal functor (F, m, m_I) between traced symmetric monoidal categories is *traced* if $F(Tr^X_{A,B}(f)) = Tr^{FX}_{FA,FB}(m; Ff; m^{-1})$ holds for any $f : A \otimes X \to B \otimes X$. □

6.2 Cyclic Sharing Models

Definition 6.2.1 (cartesian-center traced SMC)
A *cartesian-center traced symmetric monoidal category* (*cartesian-center traced SMC* for short) is a cartesian-center SMC whose symmetric monoidal category part is traced. □

Definition 6.2.2 (cartesian-center traced functors)
A *cartesian-center traced functor* between cartesian-center traced categories $\mathcal{F} : C \to S$ and $\mathcal{F}' : C' \to S'$ is a cartesian-center functor (Φ, Ψ) between them such that Ψ is traced (since Ψ is strict, we require it to preserve the trace on the nose). □

We write **CcTrSMC** for the 2-category of small cartesian-center traced symmetric monoidal categories, cartesian-center traced functors and cartesian-center natural transformations.

We again state that the simple slice construction (page 45) preserves our structure:

Lemma 6.2.3 Let $\mathcal{F} : C \to S$ be a cartesian-center traced SMC and A be an object of C (hence S). Then the Kleisli category $C//A$ of the comonad $A \times (-)$ on C is a cartesian category; the Kleisli category $S//A$ of the comonad $A \otimes (-)$ on S is a traced symmetric monoidal category; and $\mathcal{F}$ induces an identity-on-objects strict symmetric monoidal functor from $C//A$ to $S//A$ (for which we write $\mathcal{F}//A$). Therefore $\mathcal{F}//A : C//A \to S//A$ is a cartesian-center traced SMC.

Sketch of the proof: The only new point over Lemma 3.2.7 is that $S//A$ is traced. Most axioms are verified easily, except **Sliding** for which we give the calculation below. We want to show $Tr^X(f; (C \otimes g)) = Tr^Y((B \otimes g); f)$ for $f : B \otimes X \to C \otimes Y$ and $g : Y \to X$ in $S//A$, hence to show that $Tr^X((\mathcal{F}\Delta_A \otimes B \otimes X); (A \otimes f); (c_{A,C} \otimes Y); (C \otimes g))$ equals to $Tr^Y((\mathcal{F}\Delta_A \otimes B \otimes Y); (A \otimes c_{A,B} \otimes Y); (A \otimes B \otimes g); f)$ for $f : A \otimes B \otimes X \to C \otimes Y$ and $g : A \otimes Y \to X$ in S. The following picture may help with the intuition:

$$
\begin{aligned}
& Tr^X((\mathcal{F}\Delta_A \otimes B \otimes X);(A \otimes f);(c_{A,C} \otimes Y);(C \otimes g)) \\
=\ & Tr^X((\mathcal{F}\Delta_A \otimes B \otimes X);(A \otimes f);(c_{A,C} \otimes Y);(C \otimes g)) \\
=\ & Tr^{A \otimes Y}((\mathcal{F}\Delta_A \otimes B \otimes g);(A \otimes f);(c_{A,C} \otimes Y)) && \text{Sliding} \\
=\ & Tr^A(Tr^Y((\mathcal{F}\Delta_A \otimes B \otimes g);(A \otimes f);(c_{A,C} \otimes Y))) && \text{Vanishing} \\
=\ & Tr^A((\mathcal{F}\Delta_A \otimes B \otimes A);Tr^Y((A \otimes A \otimes B \otimes g);(A \otimes f));c_{A,C}) && \text{L.\& R. T.} \\
=\ & Tr^A((\mathcal{F}\Delta_A \otimes B \otimes A);(A \otimes Tr^Y((A \otimes B \otimes g);f));c_{A,C}) && \text{L.\& R. T.} \\
=\ & Tr^A((\mathcal{F}\Delta_A \otimes B \otimes A);c_{A,A \otimes B \otimes A};(Tr^Y((A \otimes B \otimes g);f) \otimes A)) \\
=\ & (\mathcal{F}\Delta_A \otimes B);Tr^A(c_{A,A \otimes B \otimes A});Tr^Y((A \otimes B \otimes g);f) && \text{L.\& R. T.} \\
=\ & (\mathcal{F}\Delta_A \otimes B);Tr^A((c_{A,A \otimes B} \otimes A);(A \otimes B \otimes c_{A,A}));Tr^Y((A \otimes B \otimes g);f) \\
=\ & (\mathcal{F}\Delta_A \otimes B);c_{A,A \otimes B};Tr^A(A \otimes B \otimes c_{A,A});Tr^Y((A \otimes B \otimes g);f) && \text{Left Tight.} \\
=\ & (\mathcal{F}\Delta_A \otimes B);c_{A,A \otimes B};(A \otimes B \otimes Tr^A(c_{A,A}));Tr^Y((A \otimes B \otimes g);f) && \text{Superpose.} \\
=\ & (\mathcal{F}\Delta_A \otimes B);c_{A,A \otimes B};Tr^Y((A \otimes B \otimes g);f) && \text{Yanking} \\
=\ & (\mathcal{F}\Delta_A \otimes B);(c_{A,A} \otimes B);(A \otimes c_{A,B});Tr^Y((A \otimes B \otimes g);f) \\
=\ & (\mathcal{F}\Delta_A \otimes B);(A \otimes c_{A,B});Tr^Y((A \otimes B \otimes g);f) \\
=\ & Tr^Y((\mathcal{F}\Delta_A \otimes B \otimes Y);(A \otimes c_{A,B} \otimes Y);(A \otimes B \otimes g);f) && \text{Left Tight.}
\end{aligned}
$$

□

Corollary 6.2.4 If C is a traced cartesian category and A is an object of C, then $C//A$ is also a traced cartesian category.   □

These results will be used in the proofs of Theorem 7.1.1 and Theorem 7.2.1.

Now we proceed to give the semantic interpretation of cyclic sharing theories in cartesian-center traced SMC's. We repeat the same pattern with Chapter 3, except the treatment of the letrec-bindings.

Definition 6.2.5 (cyclic sharing structures)
A *cyclic sharing structure* over an S-sorted signature Σ is a sharing structure (Definition 3.2.9) in a strict cartesian-center traced SMC.   □

Given a cyclic sharing structure $[\![-]\!]$ in a cartesian-center traced SMC $\mathcal{F} : C \to S$, we define $[\![x_1 : \sigma_1,\ldots,x_m : \sigma_m \vdash M : (\tau_1,\ldots,\tau_n)]\!] : [\![(\sigma_1,\ldots,\sigma_m)]\!] \to [\![(\tau_1,\ldots,\tau_n)]\!]$ in S for each well-typed term $x_1 : \sigma_1,\ldots,x_m : \sigma_m \vdash M : (\tau_1,\ldots,\tau_n)$ as follows, by induction on the typing rules.

$$\begin{aligned}
[\![\Gamma, x : \sigma \vdash x : (\sigma)]\!] &= \mathcal{F}(\pi') \\
[\![\Gamma \vdash F(M) : (\vec{\tau})]\!] &= [\![\Gamma \vdash M : (\vec{\sigma})]\!]; [\![F]\!] \\
[\![\Gamma \vdash 0 : ()]\!] &= \mathcal{F}(!) \\
[\![\Gamma \vdash M \otimes N : (\vec{\sigma}, \vec{\tau})]\!] &= \mathcal{F}(\Delta); ([\![\Gamma \vdash M : (\vec{\sigma})]\!] \otimes [\![\Gamma \vdash N : (\vec{\tau})]\!]) \\
[\![\Gamma \vdash \text{letrec } (\vec{x}) \text{ be } M \text{ in } N : (\vec{\tau})]\!] &= \\
\mathcal{F}(\Delta); (id \otimes Tr^{[\![(\vec{\sigma})]\!]} & ([\![\Gamma, \vec{x} : \vec{\sigma} \vdash M : (\vec{\sigma})]\!]; \mathcal{F}(\Delta))); [\![\Gamma, \vec{x} : \vec{\sigma} \vdash N : (\vec{\tau})]\!] \\
[\![\Gamma, x' : \sigma', x : \sigma, \Gamma' \vdash M : (\vec{\tau})]\!] &= (id \otimes \mathcal{F}(c) \otimes id); [\![\Gamma, x : \sigma, x' : \sigma', \Gamma' \vdash M : (\vec{\tau})]\!]
\end{aligned}$$

Lemma 6.2.6 Let $[\![-]\!]$ be a cyclic sharing structure. Then $[\![\vec{x} : \vec{\sigma} \vdash M : (\vec{\tau})]\!] = [\![\vec{y} : \vec{\sigma} \vdash M\{\vec{y}/\vec{x}\} : (\vec{\tau})]\!]$ where the x's and y's are disjoint.

Proof: Induction on the construction of M.   □

Definition 6.2.7 (cyclic sharing models)
A *cyclic sharing model* of a cyclic sharing theory is a cyclic sharing structure $[\![-]\!]$ in a strict cartesian-center traced SMC $\mathcal{F} : C \to S$ such that $[\![\Gamma \vdash M : (\vec{\sigma})]\!] = [\![\Gamma \vdash N : (\vec{\sigma})]\!]$ for each axiom $\Gamma \vdash M = N : (\vec{\sigma})$ of the theory.   □

Theorem 6.2.8 (soundness)
Let $[\![-]\!]$ be a sharing model of a sharing theory. If $\Gamma \vdash M = N : (\vec{\sigma})$ is derivable in the theory, then $[\![\Gamma \vdash M : (\vec{\sigma})]\!] = [\![\Gamma \vdash N : (\vec{\sigma})]\!]$.

Proof: Similar to the case of acyclic theories, but the calculation becomes considerably complicated because of the trace axioms. We shall check the axioms in Definition 2.3.3 (page 32).

- (σ_{var}): Put $p = [\![\Gamma, \vec{y} : \vec{\sigma}' \vdash z : (\sigma)]\!]$, $f = [\![\Gamma, x : \sigma, \vec{y} : \vec{\sigma}' \vdash M : (\vec{\sigma}')]\!]$ and $g = [\![\Gamma, x : \sigma, \vec{y} : \vec{\sigma}' \vdash N : (\vec{\tau})]\!]$. Also define $f' = [\![\Gamma, \vec{y} : \vec{\sigma}' \vdash M\{z/x\} : (\vec{\sigma}')]\!]$ and $g' = [\![\Gamma, \vec{y} : \vec{\sigma}' \vdash N\{z/x\} : (\vec{\tau})]\!]$. By induction on the constructions of M and N, we can show that $f' = (\mathcal{F}\Delta \otimes \mathcal{F}\Delta); (id \otimes p \otimes id); f$ and $g' = (\mathcal{F}\Delta \otimes \mathcal{F}\Delta); (id \otimes p \otimes id); g$ as in the proof of (σ_{var}) for the acyclic case. Then

$$[\![\Gamma \vdash \text{letrec } (x, \vec{y}) \text{ be } z \otimes M \text{ in } N : (\vec{\tau})]\!]$$
$$= \mathcal{F}\Delta; (id \otimes Tr^{[\![(\sigma, \vec{\sigma}')]\!]}((\mathcal{F}\Delta \otimes id \otimes \mathcal{F}\Delta); (id \otimes \mathcal{F}c \otimes id); (p \otimes f); \mathcal{F}\Delta)); g$$

$$[\![\Gamma \vdash \text{letrec } (\vec{y}) \text{ be } M\{z/x\} \text{ in } N\{z/x\} : (\vec{\tau})]\!]$$
$$= \mathcal{F}\Delta; (id \otimes Tr^{[\![\vec{\sigma}']\!]}(f'; \mathcal{F}\Delta)); g'.$$

So it suffices to show

$$Tr^{[\![(\sigma, \vec{\sigma}')]\!]}((\mathcal{F}\Delta \otimes id \otimes \mathcal{F}\Delta); (id \otimes \mathcal{F}c \otimes id); (p \otimes f); \mathcal{F}\Delta)$$
$$= \mathcal{F}\Delta; (id \otimes Tr^{[\![\vec{\sigma}']\!]}(f'; \mathcal{F}\Delta); \mathcal{F}\Delta); (p \otimes id).$$

$$\begin{aligned}
&LHS \\
&= Tr(Tr((\mathcal{F}\Delta \otimes id \otimes \mathcal{F}\Delta); (id \otimes \mathcal{F}c \otimes id); (p \otimes f); \mathcal{F}\Delta)) && \text{Vanish.} \\
&= Tr(\mathcal{F}\Delta; (p; \mathcal{F}\Delta \otimes id); (id \otimes (\mathcal{F}c \otimes id); f; \mathcal{F}\Delta)) && \text{Sld., R.T, Yank.} \\
&= Tr(\mathcal{F}\Delta; (\mathcal{F}\Delta(p \otimes p) \otimes id); (id \otimes (\mathcal{F}c \otimes id); f; \mathcal{F}\Delta)) && \\
&= RHS && \text{R.T.}
\end{aligned}$$

- (id) is the same as in the acyclic case.

- (ass$_1$): Put $f = [\![\Gamma, \vec{x} : \vec{\sigma}, \vec{y} : \vec{\sigma}' \vdash L : (\vec{\sigma}')]\!]$, $g = [\![\Gamma, \vec{x} : \vec{\sigma}, \vec{y} : \vec{\sigma}' \vdash M : (\vec{\sigma})]\!]$ and
  $h = [\![\Gamma, \vec{x} : \vec{\sigma} \vdash N : (\vec{\tau})]\!]$. Then

$$[\![\Gamma \vdash \text{letrec } (\vec{x}) \text{ be } (\text{letrec } (\vec{y}) \text{ be } L \text{ in } M) \text{ in } N : (\vec{\tau})]\!]$$
$$= \quad \mathcal{F}\Delta; (id \otimes Tr^{[\![(\vec{\sigma})]\!]}(\mathcal{F}\Delta; (id \otimes Tr^{[\![(\vec{\sigma}')]\!]}(f; \mathcal{F}\Delta)); g; \mathcal{F}\Delta)); h$$

$$[\![\Gamma \vdash \text{letrec } (\vec{x}, \vec{y}) \text{ be } M \otimes L \text{ in } N :: (\vec{\tau})]\!]$$
$$= \quad \mathcal{F}\Delta; (id \otimes Tr^{[\![(\vec{\sigma}, \vec{\sigma}')]\!]}(\mathcal{F}\Delta; (g \otimes f); \mathcal{F}\Delta)); \mathcal{F}\pi; h.$$

We shall show

$$Tr^{[\![(\vec{\sigma})]\!]}(\mathcal{F}\Delta; (id \otimes Tr^{[\![(\vec{\sigma}')]\!]}(f; \mathcal{F}\Delta)); g; \mathcal{F}\Delta) = Tr^{[\![(\vec{\sigma}, \vec{\sigma}')]\!]}(\mathcal{F}\Delta; (g \otimes f); \mathcal{F}\Delta); \mathcal{F}\pi.$$

	RHS	
$=$	$Tr(Tr(\mathcal{F}\Delta; (g \otimes f); \mathcal{F}\Delta)); \mathcal{F}\pi$	Vanishing
$=$	$Tr(\mathcal{F}\Delta; Tr((id \otimes \mathcal{F}\Delta); (id \otimes \mathcal{F}c \otimes id); (id \otimes f; \mathcal{F}\Delta));$	
	$(g; \mathcal{F}\Delta \otimes id)); \pi$	L.T.&R.T.
$=$	$Tr(\mathcal{F}\Delta; (id \otimes Tr((id \otimes \mathcal{F}\Delta); (\mathcal{F}c \otimes id); (id \otimes f; \mathcal{F}\Delta)));$	
	$(g; \mathcal{F}\Delta \otimes id); \pi$	Superposing
$=$	$Tr(\mathcal{F}\Delta; (id \otimes Tr((id \otimes \mathcal{F}\Delta); (\mathcal{F}c \otimes id); (id \otimes f))); g; \mathcal{F}\Delta)$	R.T.
$=$	$Tr(\mathcal{F}\Delta; (id \otimes Tr((\mathcal{F}c \otimes id); (id \otimes f); (id \otimes \mathcal{F}\Delta))); g; \mathcal{F}\Delta)$	Sliding
$=$	LHS	L.T.&Yank.

- (ass$_2$) is the same as in the acyclic theory.

- $\otimes_i$ are the same as in the acyclic theory.

- (perm): Put $f = [\![\Gamma, \vec{x} : \vec{\sigma}, \vec{y} : \vec{\sigma}', \vec{z} : \vec{\sigma}'' \vdash M_1 \otimes M_2 \otimes M_3 : (\vec{\sigma}, \vec{\sigma}', \vec{\sigma}'')]\!]$ and $g = [\![\Gamma, \vec{y} : \vec{\sigma}', \vec{x} : \vec{\sigma}, \vec{z} : \vec{\sigma}'' \vdash M_2 \otimes M_1 \otimes M_3 : (\vec{\sigma}', \vec{\sigma}, \vec{\sigma}'')]\!]$. Then $f = (id \otimes \mathcal{F}c \otimes id); g; (\mathcal{F}c \otimes id)$, and by Sliding and Right Tightening $Tr^{[\![(\vec{\sigma}, \vec{\sigma}', \vec{\sigma}'')]\!]}(f; \mathcal{F}\Delta) = Tr^{[\![(\vec{\sigma}', \vec{\sigma}, \vec{\sigma}'')]\!]}(g; \mathcal{F}\Delta); (\mathcal{F}c \otimes id)$. Now (perm) is proved by

$$[\![\Gamma \vdash \text{letrec } (\vec{x}, \vec{y}, \vec{z}) \text{ be } M_1 \otimes M_2 \otimes M_3 \text{ in } N : (\vec{\tau})]\!]$$
$$= \quad \mathcal{F}\Delta; (id \otimes Tr^{[\![(\vec{\sigma}, \vec{\sigma}', \vec{\sigma}'')]\!]}(f; \mathcal{F}\Delta)); [\![\Gamma, \vec{x} : \vec{\sigma}, \vec{y} : \vec{\sigma}', \vec{z} : \vec{\sigma}'' \vdash N : (\vec{\tau})]\!]$$
$$= \quad \mathcal{F}\Delta; (id \otimes Tr^{[\![(\vec{\sigma}', \vec{\sigma}, \vec{\sigma}'')]\!]}(g; \mathcal{F}\Delta); (\mathcal{F}c \otimes id)); [\![\Gamma, \vec{x} : \vec{\sigma}, \vec{y} : \vec{\sigma}', \vec{z} : \vec{\sigma}'' \vdash N : (\vec{\tau})]\!]$$
$$= \quad \mathcal{F}\Delta; (id \otimes Tr^{[\![(\vec{\sigma}', \vec{\sigma}, \vec{\sigma}'')]\!]}(g; \mathcal{F}\Delta)); [\![\Gamma, \vec{y} : \vec{\sigma}', \vec{x} : \vec{\sigma}, \vec{z} : \vec{\sigma}'' \vdash N : (\vec{\tau})]\!]$$
$$= \quad [\![\Gamma \vdash \text{letrec } (\vec{y}, \vec{x}, \vec{z}) \text{ be } M_2 \otimes M_1 \otimes M_3 \text{ in } N : (\vec{\tau})]\!]$$

- (subst) is the same as in the acyclic case.

$\square$

For a cyclic sharing theory $\mathbb{T}$ (determined by a set of axioms) over a signature Σ, we write **CSharingMod**$(\mathbb{T}, (\mathcal{F} : C \to S))$ for the category of $\mathbb{T}$'s cyclic sharing models in a cartesian-center traced SMC $\mathcal{F} : C \to S$ and the homomorphisms between models.

6.3 The Classifying Category

Proposition 6.3.1 Given a cyclic sharing theory $\mathbb{T}$ over Σ, there is a strict cartesian-center traced SMC $\mathcal{F}_{\mathbb{T}} : C_{\mathbb{T}} \to \mathcal{S}_{\mathbb{T}}$ obtained exactly in the same way as we did for constructing the classifying category of an acyclic sharing theory. The only difference is that let-bindings are replaced by letrec-bindings

$$[\Gamma \vdash M : (\vec{\sigma})]; [\vec{x} : \vec{\sigma} \vdash N : (\vec{\tau})] = [\Gamma \vdash \text{letrec } (\vec{x}) \text{ be } M \text{ in } N : (\vec{\tau})]$$

which are powerful enough to determine a trace structure on $\mathcal{S}_{\mathbb{T}}$ as

$$Tr^{(\vec{\sigma})}([\Gamma, \vec{x} : \vec{\sigma} \vdash M : (\vec{\tau}, \vec{\sigma})]) = [\Gamma \vdash \text{letrec } (\vec{y}, \vec{x}) \text{ be } M \text{ in } \vec{y} : (\vec{\tau})].$$

<u>Proof:</u> We need to verify the axioms for traces.

- Vanishing:

$$
\begin{aligned}
&Tr^{(\vec{\sigma})}(Tr^{(\vec{\sigma}')}([\Gamma, \vec{x} : \vec{\sigma}, \vec{x}' : \vec{\sigma}' \vdash M : (\vec{\tau}, \vec{\sigma}, \vec{\sigma}')])) \\
=\ &Tr^{(\vec{\sigma})}([\Gamma, \vec{x} : \vec{\sigma} \vdash \text{ letrec } (\vec{y}, \vec{z}, \vec{x}') \text{ be } M \text{ in } \vec{y} \otimes \vec{z} : (\vec{\tau}, \vec{\sigma})]) \\
=\ &[\Gamma \vdash \text{ letrec } (\vec{y}', \vec{x}) \text{ be } (\text{letrec } (\vec{y}, \vec{z}, \vec{x}') \text{ be } M \text{ in } \vec{y} \otimes \vec{z}) \text{ in } \vec{y}' : (\vec{\tau})] \\
=\ &[\Gamma \vdash \text{ letrec } (\vec{y}', \vec{x}, \vec{y}, \vec{z}, \vec{x}') \text{ be } \vec{y} \otimes \vec{z} \otimes M \text{ in } \vec{y}' : (\vec{\tau})] && (\text{ass}_1) \\
=\ &[\Gamma \vdash \text{ letrec } (\vec{x}, \vec{y}, \vec{z}, \vec{x}') \text{ be } \vec{z} \otimes M \text{ in } \vec{y} : (\vec{\tau})] && (\sigma_{\text{var}}) \\
=\ &[\Gamma \vdash \text{ letrec } (\vec{y}, \vec{z}, \vec{x}') \text{ be } M\{\vec{z}/\vec{x}\} \text{ in } \vec{y} : (\vec{\tau})] && (\sigma_{\text{var}}) \\
=\ &[\Gamma \vdash \text{ letrec } (\vec{y}, \vec{x}, \vec{x}') \text{ be } M \text{ in } \vec{y} : (\vec{\tau})] && (\alpha) \\
=\ &Tr^{(\vec{\sigma}, \vec{\sigma}')}([\Gamma, \vec{x} : \vec{\sigma}, \vec{x}' : \vec{\sigma}' \vdash M : (\vec{\tau}, \vec{\sigma}, \vec{\sigma}')])
\end{aligned}
$$

The other vanishing axiom is trivial.

- Superposing:

$$
\begin{aligned}
&Tr^{(\vec{\sigma})}(id_{(\tau)} \otimes [\Gamma, \vec{x} : \vec{\sigma} \vdash M : (\vec{\sigma}', \vec{\sigma})]) \\
=\ &Tr^{(\vec{\sigma})}([\vec{y} : \vec{\tau}, \Gamma, \vec{x} : \vec{\sigma} \vdash \vec{y} \otimes M : (\vec{y}, \vec{\sigma}', \vec{\sigma})]) \\
=\ &[\vec{y} : \vec{\tau}, \Gamma \vdash \text{ letrec } (\vec{y}', \vec{x}', \vec{x}) \text{ be } \vec{y} \otimes M \text{ in } \vec{y}' \otimes \vec{x}' : (\vec{\tau}, \vec{\sigma}')] \\
=\ &[\vec{y} : \vec{\tau}, \Gamma \vdash \text{ letrec } (\vec{x}', \vec{x}) \text{ be } M \text{ in } \vec{y} \otimes \vec{x}' : (\vec{\tau}, \vec{\sigma}')] && (\sigma_{\text{var}}) \\
=\ &[\vec{y} : \vec{\tau}, \Gamma \vdash \vec{y} \otimes (\text{letrec } (\vec{x}', \vec{x}) \text{ be } M \text{ in } \vec{x}') : (\vec{\tau}, \vec{\sigma}')] && (\otimes_1) \\
=\ &id_{(\vec{\tau})} \otimes [\Gamma \vdash \text{ letrec } (\vec{x}', \vec{x}) \text{ be } M \text{ in } \vec{x}' : (\vec{\sigma}')] \\
=\ &id_{(\vec{\tau})} \otimes Tr^{(\vec{\sigma})}([\Gamma, \vec{x} : \vec{\sigma} \vdash M : (\vec{\sigma}', \vec{\sigma})])
\end{aligned}
$$

- Yanking:

$$
\begin{aligned}
&Tr^{(\vec{\sigma})}(c_{(\vec{\sigma}),(\vec{\sigma})}) \\
=\ &Tr^{(\vec{\sigma})}([\vec{x} : \vec{\sigma}, \vec{y} : \vec{\sigma} \vdash \vec{y} \otimes \vec{x} : (\vec{\sigma}, \vec{\sigma})]) \\
=\ &[\vec{x} : \vec{\sigma} \vdash \text{ letrec } (\vec{z}, \vec{y}) \text{ be } \vec{y} \otimes \vec{x} \text{ in } \vec{z} : (\vec{\sigma})] \\
=\ &[\vec{x} : \vec{\sigma} \vdash \text{ letrec } (\vec{y}) \text{ be } \vec{x} \text{ in } \vec{y} : (\vec{\sigma})] && (\sigma_{\text{var}}) \\
=\ &[\vec{x} : \vec{\sigma} \vdash \vec{x} : (\vec{\sigma})] && (\text{id}) \\
=\ &id_{(\vec{\sigma})}
\end{aligned}
$$

- Left Tightening:

$$
\begin{aligned}
&\quad Tr^{(\vec{\sigma})}(([\Gamma \vdash M : (\vec{\tau})] \otimes id_{(\vec{\sigma})}); [\vec{y} : \vec{\tau}, \vec{x} : \vec{\sigma} \vdash N : (\vec{\sigma}', \vec{\sigma})]) \\
&= Tr^{(\vec{\sigma})}([\Gamma, \vec{z} : \vec{\sigma} \vdash M \otimes \vec{z} : (\vec{\tau}, \vec{\sigma})]; [\vec{y} : \vec{\tau}, \vec{x} : \vec{\sigma} \vdash N : (\vec{\sigma}', \vec{\sigma})]) \\
&= Tr^{(\vec{\sigma})}([\Gamma, \vec{z} : \vec{\sigma} \vdash \text{letrec } (\vec{y}, \vec{x}) \text{ be } M \otimes \vec{z} \text{ in } N : (\vec{\sigma}', \vec{\sigma})]) \\
&= [\Gamma \vdash \text{letrec } (\vec{u}, \vec{z}) \text{ be } (\text{letrec } (\vec{y}, \vec{x}) \text{ be } M \otimes \vec{z} \text{ in } N) \text{ in } \vec{u} : (\vec{\sigma}')] \\
&= [\Gamma \vdash \text{letrec } (\vec{u}, \vec{z}, \vec{y}, \vec{x}) \text{ be } N \otimes M \otimes \vec{z} \text{ in } \vec{u} : (\vec{\sigma}')] && (\text{ass}_1) \\
&= [\Gamma \vdash \text{letrec } (\vec{x}, \vec{u}, \vec{z}, \vec{y}) \text{ be } \vec{z} \otimes N \otimes M \text{ in } \vec{u} : (\vec{\sigma}')] && (\text{perm}) \\
&= [\Gamma \vdash \text{letrec } (\vec{u}, \vec{z}, \vec{y}) \text{ be } N\{\vec{z}/\vec{x}\} \otimes M \text{ in } \vec{u} : (\vec{\sigma}')] && (\sigma_{\text{var}}) \\
&= [\Gamma \vdash \text{letrec } (\vec{y}, \vec{u}, \vec{z}) \text{ be } M \otimes N\{\vec{z}/\vec{x}\} \text{ in } \vec{u} : (\vec{\sigma}')] && (\text{perm}) \\
&= [\Gamma \vdash \text{letrec } (\vec{y}) \text{ be } M \text{ in letrec } (\vec{u}, \vec{z}) \text{ be } N\{\vec{z}/\vec{x}\} \text{ in } \vec{u} : (\vec{\sigma}')] && (\text{ass}_2) \\
&= [\Gamma \vdash \text{letrec } (\vec{y}) \text{ be } M \text{ in letrec } (\vec{u}, \vec{x}) \text{ be } N \text{ in } \vec{u} : (\vec{\sigma}')] && (\alpha) \\
&= [\Gamma \vdash M : (\vec{\tau})]; [\vec{y} : \vec{\tau} \vdash \text{letrec } (\vec{u}, \vec{x}) \text{ be } N \text{ in } \vec{u} : (\vec{\sigma}')] \\
&= [\Gamma \vdash M : (\vec{\tau})]; Tr^{(\vec{\sigma})}([\vec{y} : \vec{\tau}, \vec{x} : \vec{\sigma} \vdash N : (\vec{\sigma}', \vec{\sigma})])
\end{aligned}
$$

- Right Tightening:

$$
\begin{aligned}
&\quad Tr^{(\vec{\sigma})}([\Gamma, \vec{x} : \vec{\sigma} \vdash M : (\vec{\tau}, \vec{\sigma})]; ([\vec{y} : \vec{\tau} \vdash N : (\vec{\tau}')] \otimes id_{(\vec{\sigma})})) \\
&= Tr^{(\vec{\sigma})}([\Gamma, \vec{x} : \vec{\sigma} \vdash M : (\vec{\tau}, \vec{\sigma})]; [\vec{y} : \vec{\tau}, \vec{z} : \vec{\sigma} \vdash N \otimes \vec{z} : (\vec{\tau}', \vec{\sigma})]) \\
&= Tr^{(\vec{\sigma})}([\Gamma, \vec{x} : \vec{\sigma} \vdash \text{letrec } (\vec{y}, \vec{z}) \text{ be } M \text{ in } N \otimes \vec{z} : (\vec{\tau}', \vec{\sigma})]) \\
&= [\Gamma \vdash \text{letrec } (\vec{u}, \vec{x}) \text{ be } (\text{letrec } (\vec{y}, \vec{z}) \text{ be } M \text{ in } N \otimes \vec{z}) \text{ in } \vec{u} : (\vec{\tau}')] \\
&= [\Gamma \vdash \text{letrec } (\vec{u}, \vec{x}, \vec{y}, \vec{z}) \text{ be } N \otimes \vec{z} \otimes M \text{ in } \vec{u} : (\vec{\tau}')] && (\text{ass}_1) \\
&= [\Gamma \vdash \text{letrec } (\vec{x}, \vec{y}, \vec{z}, \vec{u}) \text{ be } \vec{z} \otimes M \otimes N \text{ in } \vec{u} : (\vec{\tau}')] && (\text{perm}) \\
&= [\Gamma \vdash \text{letrec } (\vec{y}, \vec{z}, \vec{u}) \text{ be } M \otimes N\{\vec{z}/\vec{x}\} \text{ in } \vec{u} : (\vec{\tau}')] && (\sigma_{\text{var}}) \\
&= [\Gamma \vdash \text{letrec } (\vec{y}, \vec{x}, \vec{u}) \text{ be } M \otimes N \text{ in } \vec{u} : (\vec{\tau}')] && (\alpha) \\
&= [\Gamma \vdash \text{letrec } (\vec{y}, \vec{x}) \text{ be } M \text{ in letrec } (\vec{u}) \text{ be } N \text{ in } \vec{u} : (\vec{\tau}')] && (\text{ass}_2) \\
&= [\Gamma \vdash \text{letrec } (\vec{y}, \vec{x}) \text{ be } M \text{ in } N : (\vec{\tau}')] && (\text{id}) \\
&= [\Gamma \vdash \text{letrec } (\vec{v}, \vec{x}) \text{ be } M \text{ in } N\{\vec{v}/\vec{y}\} : (\vec{\tau}')] && (\alpha) \\
&= [\Gamma \vdash \text{letrec } (\vec{y}, \vec{v}, \vec{x}) \text{ be } \vec{v} \otimes M \text{ in } N : (\vec{\tau}')] && (\sigma_{\text{var}}) \\
&= [\Gamma \vdash \text{letrec } (\vec{y}) \text{ be } (\text{letrec } (\vec{v}, \vec{x}) \text{ be } M \text{ in } \vec{v}) \text{ in } N : (\vec{\tau}')] && (\text{ass}_1) \\
&= [\Gamma \vdash \text{letrec } (\vec{v}, \vec{x}) \text{ be } M \text{ in } \vec{v} : (\vec{\tau})]; [\vec{y} : \vec{\tau} \vdash N : (\vec{\tau}')] \\
&= Tr^{(\vec{\sigma})}([\Gamma, \vec{x} : \vec{\sigma} \vdash M : (\vec{\tau}, \vec{\sigma})]); [\vec{y} : \vec{\tau} \vdash N : (\vec{\tau}')]
\end{aligned}
$$

- Sliding:

$$
\begin{aligned}
&\quad Tr^{(\vec{\delta})}([\vec{y}:\vec{\tau},\vec{x}:\vec{\sigma} \vdash M : (\vec{\tau}',\vec{\sigma}')]; (id_{(\vec{\tau}')} \otimes [\vec{z}:\vec{\delta}' \vdash N : (\vec{\delta})])) \\
&= Tr^{(\vec{\delta})}([\vec{y}:\vec{\tau},\vec{x}:\vec{\sigma} \vdash M : (\vec{\tau}',\vec{\sigma}')]; [\vec{u}:\vec{\tau}',\vec{z}:\vec{\delta}' \vdash \vec{u} \otimes N : (\vec{\tau}',\vec{\delta})]) \\
&= Tr^{(\vec{\delta})}([\vec{y}:\vec{\tau},\vec{x}:\vec{\sigma} \vdash \text{letrec } (\vec{u},\vec{z}) \text{ be } M \text{ in } \vec{u} \otimes N : (\vec{\tau}',\vec{\delta})]) \\
&= [\vec{y}:\vec{\tau} \vdash \text{letrec } (\vec{v},\vec{x}) \text{ be } (\text{letrec } (\vec{u},\vec{z}) \text{ be } M \text{ in } \vec{u} \otimes N) \text{ in } \vec{v} : (\vec{\tau}')] \\
&= [\vec{y}:\vec{\tau} \vdash \text{letrec } (\vec{v},\vec{x},\vec{u},\vec{z}) \text{ be } \vec{u} \otimes N \otimes M \text{ in } \vec{v} : (\vec{\tau}')] && (\text{ass}_1) \\
&= [\vec{y}:\vec{\tau} \vdash \text{letrec } (\vec{x},\vec{u},\vec{z}) \text{ be } N \otimes M \text{ in } \vec{u} : (\vec{\tau}')] && (\sigma_{var}) \\
&= [\vec{w}:\vec{\tau} \vdash \text{letrec } (\vec{x},\vec{s},\vec{z}) \text{ be } N \otimes M\{\vec{w}/\vec{y}\} \text{ in } \vec{s} : (\vec{\tau}')] && (\alpha) \\
&= [\vec{w}:\vec{\tau} \vdash \text{letrec } (\vec{y},\vec{x},\vec{s},\vec{z}) \text{ be } \vec{w} \otimes N \otimes M \text{ in } \vec{s} : (\vec{\tau}')] && (\sigma_{var}) \\
&= [\vec{w}:\vec{\tau} \vdash \text{letrec } (\vec{s},\vec{z},\vec{y},\vec{x}) \text{ be } M \otimes \vec{w} \otimes N \text{ in } \vec{s} : (\vec{\tau}')] && (\text{perm}) \\
&= [\vec{w}:\vec{\tau} \vdash \text{letrec } (\vec{s},\vec{z}) \text{ be } (\text{letrec } (\vec{y},\vec{x}) \text{ be } \vec{w} \otimes N \text{ in } M) \text{ in } \vec{s} : (\vec{\tau}')] && (\text{ass}_1) \\
&= Tr^{(\vec{\delta}')}([\vec{w}:\vec{\tau},\vec{z}:\vec{\delta}' \vdash \text{letrec } (\vec{y},\vec{x}) \text{ be } \vec{w} \otimes N \text{ in } M : (\vec{\tau}',\vec{\sigma}')]) \\
&= Tr^{(\vec{\delta}')}([\vec{w}:\vec{\tau},\vec{z}:\vec{\delta}' \vdash \vec{w} \otimes N : (\vec{\tau},\vec{\delta})]; [\vec{y}:\vec{\tau},\vec{x}:\vec{\sigma} \vdash M : (\vec{\tau}',\vec{\sigma}')]) \\
&= Tr^{(\vec{\delta}')}((id_{(\vec{\tau})} \otimes [\vec{z}:\vec{\delta}' \vdash N : (\vec{\delta})]); [\vec{y}:\vec{\tau},\vec{x}:\vec{\sigma} \vdash M : (\vec{\tau}',\vec{\sigma}')])
\end{aligned}
$$

$\square$

Theorem 6.3.2 (completeness)
Given a theory $\mathbb{T}$, there is a complete model in $\mathcal{F}_{\mathbb{T}} : C_{\mathbb{T}} \to S_{\mathbb{T}}$, given by $[\![\sigma]\!] = \sigma$ and
$[\![F]\!] = [\vec{x}:\vec{\sigma} \vdash F(\vec{x}) : (\vec{\tau})]$ for $F : (\vec{\sigma}) \to (\vec{\tau})$. $\square$

Theorem 6.3.3

$$\mathbf{CcTrSMC}((C_{\mathbb{T}} \xrightarrow{\mathcal{F}_{\mathbb{T}}} S_{\mathbb{T}}),(C \xrightarrow{\mathcal{F}} S)) \simeq \mathbf{CSharingMod}(\mathbb{T},(C \xrightarrow{\mathcal{F}} S)).$$

$\square$

7
Recursion from Cyclic Sharing

We have studied models of both higher-order acyclic sharing theories (Chapter 4) and (first-order) cyclic sharing theories (Chapter 6), as two orthogonal extensions of (first-order) acyclic sharing theories (Chapter 3). Now we are ready to deal with the most interesting setting in this thesis: the combination of the higher-order extension with cyclic sharing. The consequence is more than an ad hoc combination of two ideas. First, we get a wider class of models of recursive computation, which properly contains the traditional models such as domain theoretic examples but also some non-traditional examples, which are suitable for explaining recursive computation from resource-sensitive cyclic computation, i.e. *recursion from cyclic sharing*. Second, our theory is closely related to *cyclic lambda calculi* (lambda calculi with cyclic sharing) which have been studied as a foundation for practical implementation of functional recursive computation. Therefore our result relates a rich class of models and practically interesting calculi: cyclic lambda calculi serve as languages for our new class of models of recursion, while our models serve as a semantic counterpart of cyclic lambda calculi.

Unlike the previous chapters, we begin with semantic observations, rather than starting from the syntax (which is more or less derivable from the previous ones anyway). We first show a surprisingly simple connection between fixed point operators and traces on cartesian categories. This observation itself stays in the traditional contexts without a notion of sharing, but we then extend this result to cartesian centrally closed traced SMC's, thus our models of higher-order cyclic sharing. After that we introduce higher-order cyclic sharing theories, paying some attention to the related systems known as cyclic lambda calculi. As an application, we analyze fixed point operators definable in cyclic lambda calculi using our semantic models.

Some results in this chapter are also reported in [38].

7.1 Fixed Points in Traced Cartesian Categories

Compact closed categories whose monoidal product is cartesian are trivial. This is not the case for traced categories. In fact, in [50] it is shown that the category of sets and binary relations with its biproduct as the monoidal product is traced. Actually we find traced cartesian categories interesting in the context of semantics for recursive computation. Here is a theorem to relate the traditional fixed point operators and traces on cartesian categories, proved by Martin Hyland and the author independently:

Theorem 7.1.1 A cartesian category C is traced if and only if it has a family of functions

$$(-)^{\dagger_{A,X}} : C(A \times X, X) \longrightarrow C(A, X)$$

(in below, parameters A, X may be omitted) such that

1. $(-)^{\dagger}$ is a *parametrized fixed point operator*; for $f : A \times X \longrightarrow X$, $f^{\dagger} : A \longrightarrow X$ satisfies $f^{\dagger} = \langle id_A, f^{\dagger} \rangle; f$.

2. $(-)^{\dagger}$ is natural in A; for $f : A \times X \longrightarrow X$ and $g : B \longrightarrow A$, $((g \times id_X); f)^{\dagger} = g; f^{\dagger} : B \longrightarrow X$.

3. $(-)^{\dagger}$ is natural in X; for $f : A \times X \longrightarrow Y$ and $g : Y \longrightarrow X$, $(f;g)^{\dagger} = ((id_A \times g); f)^{\dagger}; g : A \longrightarrow X$.

4. $(-)^{\dagger}$ satisfies *Bekič's lemma*; for $f : A \times X \times Y \longrightarrow X$ and $g : A \times X \times Y \longrightarrow Y$, $\langle f, g \rangle^{\dagger} = \langle id_A, ((\langle id_{A \times X}, g^{\dagger} \rangle; f)^{\dagger}) \rangle; \langle \pi'_{A,X}, g^{\dagger} \rangle : A \longrightarrow X \times Y$.

Sketch of the proof: (The full calculation is found in Appendix A.) From a trace operator Tr, we define a fixed point operator $(-)^{\dagger}$ by

$$f^{\dagger} = Tr^X(f; \Delta_X) : A \longrightarrow X$$

for $f : A \times X \longrightarrow X$. Conversely, from a fixed point operator $(-)^{\dagger}$ we define a trace Tr by

$$Tr^X(f) = \langle id_A, (f; \pi'_{B,X})^{\dagger} \rangle; f; \pi_{B,X} : A \longrightarrow B$$

(equivalently $((id_A \times \pi'_{B,X}); f)^{\dagger}; \pi_{B,X})$ for $f : A \times X \longrightarrow B \times X$. We note that these constructions are mutually inverse. $\qquad\square$

There are several equivalent formulations of this result. For instance, in the presence of other conditions, we can restrict 3 to the case that g is a symmetry (c.f. Lemma 1.1. of [50]). For another – practically useful – example, Hyland has shown that axioms 1~4 are equivalent to 2 and

- *(parametrized) dinaturality*: $(\langle \pi_{A,X}, f \rangle; g)^{\dagger} = \langle id_A, (\langle \pi_{A,Y}, g \rangle; f)^{\dagger} \rangle; g : A \longrightarrow X$ for $f : A \times X \longrightarrow Y$ and $g : A \times Y \longrightarrow X$

- *diagonal property*: $(f^{\dagger})^{\dagger} = ((id_A \times \langle id_X, id_X \rangle); f)^{\dagger}$ for $f : A \times X \times X \longrightarrow X$.

This axiomatization is the same as that of "Conway cartesian categories" in [22]. Further variations are: 2,4 with dinaturality; and 1,2,4 with the symmetric form of 4.

Perhaps the simplest example is the opposite of the category of sets and partial functions with coproduct as the monoidal product; the trace is given by a form of *feedback* which maps a partial function $f : X \to A + X$ to $f^{\dagger} : X \to A$, determined by iterating f until we get an answer in A if it exists. Similar settings are studied in detail in [21].

An immediate consequence of Theorem 7.1.1 is the close relationship between traces and the least fixed point operators in traditional domain theory.

Example 7.1.2 (the least fixed point operator on domains)
Consider the cartesian closed category **Dom** of Scott domains and continuous functions. The least fixed point operator satisfies conditions 1~4, thus determines a trace operator given by $Tr^X(f) = \lambda a^A.\pi(f(a, \bigcup^n(\lambda x^X.\pi'(f(a,x)))^n(\perp_X)))$
$: A \to B$ for $f : A \times X \to B \times X$. Since the least fixed point operator is the unique dinatural fixed point operator on **Dom**, the trace above is the unique one on **Dom**. □

The same is true for several cartesian closed categories arising from domain theory. In fact, a systematic account is possible. Simpson [82] has shown that, under a mild condition, in cartesian closed full subcategories of the category of algebraic cpo's, the least fixed point operator is characterized as the unique dinatural fixed point operator. On the other hand, it is easy to see that the least fixed point operators satisfy the conditions of Theorem 7.1.1. Therefore, in many such categories, a trace uniquely exists and is determined by the least fixed point operator. However, we note that there are at least two traces in the category of continuous lattices, an instance which does not satisfy Simpson's condition; this category has two fixed point operators which satisfy our conditions – the least one and the greatest continuous one.

Further justification of our axiomatization of fixed point operators comes from recent work on *axiomatic domain theory* which provides a more abstract and systematic treatment of domains and covers a wider range of models of domain theory than the traditional order-theoretic approach. For this, we assume some working knowledge of this topic as found in [81]. Readers who are not familiar with this topic may skip to the next section.

Example 7.1.3 (axiomatic domain theory)
Consider a cartesian closed category C (category of "predomains") equipped with a commutative monad L (the "lift") such that the Kleisli category C_L (category of predomains and "partial maps") is *algebraically compact* [32]. This setting provides a canonical fixed point operator (derived from the *fixpoint object* [27]) on the category of "domains" (obtained as the Kleisli category of the induced comonad on the Eilenberg-Moore category C^L) which satisfies our axioms – Bekič's lemma is proved from the algebraic compactness of C_L [73] (this idea is due to Plotkin). Thus the requirement for solving recursive domain equations (algebraic compactness) implies that the resulting category of domains is traced. □

Regarding these facts, we believe that traces provide a good characterization of fixed point operators in traditional denotational semantics.

We conclude this section by observing an attractive fact which suggests how natural our trace-fixpoint correspondence is. Our correspondence preserves a fundamental concept on fixed point operators called *uniformity*, also known as Plotkin's condition. This is important because fixed point operators are often canonically and uniquely characterized by this property.

Proposition 7.1.4 In a traced cartesian category, the following two conditions are equivalent for any $h : X \longrightarrow Y$.

- (Uniformity of the trace operator) For any f and g,

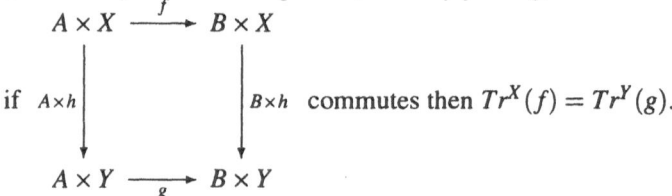

if $A \times h$ ... $B \times h$ commutes then $Tr^X(f) = Tr^Y(g)$.

- (Uniformity of the fixed point operator) For any f and g,

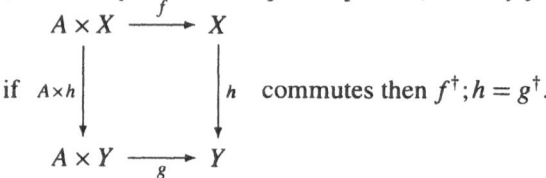

if $A \times h$ ... h commutes then $f^\dagger; h = g^\dagger$.

<u>Proof:</u> See Appendix.   □

In the case of domain-theoretic categories, the second condition is equivalent to saying that h is a strict map ($\perp$-preserving map). This fact suggests the possibility of studying the notion of strict maps and uniformity of fixed points in more general settings as in the following section. In particular, the first condition makes sense in any traced monoidal category. We remark that this notion of strict maps seems to be far more flexible (or fragile) than that we have in traditional domain theory; in some traced monoidal categories every arrow is strict (e.g. category of sets and partial functions), whereas in some cases only isomorphisms are strict (e.g. category of sets and relations).

Remark 7.1.5 Here is an interesting unanswered question: when do strict maps (defined with respect to the uniformity condition on fixed point operators or traces as above) form a category? Though in many concrete examples (including all domain theoretic examples) strict maps do compose, our abstract definition is not strong enough to ensure that.

More generally, it is straightforward to define the notion of "uniform transformations" (with respect to a class of "strict maps") between mixed variant functors in a similar manner to dinatural transformations [59, 85]. Unlike dinaturals, uniform transformations do compose, but the tradeoff is the failure of the compositionality of (abstract) strict maps. We are not keen to go further in such a general setting, but this question may be a key to develop a theory of recursive computation based on traced categories – we admit that our understanding on this topic is still premature.   □

7.2   Generalized Fixed Points

Our observation so far says that to have an abstract trace is to have a fixed point operator in the traditional sense, provided the monoidal product is cartesian. However, regarding our motivation to model cyclic sharing, this setting is somewhat restrictive

– in a cartesian category (regarded as an algebraic theory) arbitrary substitution is justified, thus there is no non-trivial notion of sharing.

Our next step is then to extend this result to our models of cyclic sharing graphs. It turns out that the combination of the higher-order extension with the cyclic sharing theories is sufficient for supporting a generalization of the result of the last section. The essential point is the use of an adjunction between a cartesian category and a traced monoidal category for creating recursion, which is spelled out below.

Let $\mathcal{F} : C \longrightarrow S$ be an identity-on-objects strict symmetric monoidal functor from a cartesian category C to a traced symmetric monoidal category S, with a right adjoint. That is, $\mathcal{F} : C \longrightarrow S$ is a cartesian-center traced SMC with a right adjoint.

Theorem 7.2.1 Given $\mathcal{F} : C \longrightarrow S$ as above, there is a family of functions

$$(-)^{\dagger_{A,X}} : S(A \otimes X, X) \longrightarrow S(A, X)$$

such that

1. $(-)^{\dagger}$ is a parametrized dinatural fixed point operator: for $f : A \otimes X \longrightarrow Y$ in S and $g : A \otimes Y \longrightarrow X$ in S, $((\mathcal{F}(\Delta_A) \otimes id_X); (id_A \otimes f); g)^{\dagger} = \mathcal{F}(\Delta_A); (id_A \otimes ((\mathcal{F}(\Delta_A) \otimes id_Y); (id_A \otimes g); f)^{\dagger}); g : A \longrightarrow X$.

2. $(-)^{\dagger}$ is natural in A in C; for $f : A \otimes X \longrightarrow X$ in S and $g : B \longrightarrow A$ in C, $((\mathcal{F}(g) \otimes id_X); f)^{\dagger} = \mathcal{F}(g); f^{\dagger} : B \longrightarrow X$.

Sketch of the proof: (The full calculation is found in Appendix A.) Let us write $U : S \longrightarrow C$ for the right adjoint of $\mathcal{F}$, and $\varepsilon_X : UX \longrightarrow X$ (in S) for the counit. By definition, we have a natural isomorphism $(-)^* : S(A, B) \xrightarrow{\sim} C(A, UB)$. We also define $\theta_{A,X} : A \times UX \longrightarrow U(A \otimes X)$ in C by $\theta_{A,X} = (id_A \otimes \varepsilon_X)^*$. Now we define $(-)^{\dagger}$ by

$$f^{\dagger} = Tr^{UX}(\mathcal{F}(\theta_{A,X}; Uf; \Delta_{UX})); \varepsilon_X : A \longrightarrow X \quad \text{in } S$$

for $f : A \otimes X \longrightarrow X$ in S. □

The statement above is slightly stronger than Theorem 3.5 in [38]. The first condition immediately implies that $(-)^{\dagger}$ is a parametrized fixed point operator, in the sense that, for $f : A \otimes X \longrightarrow X$ in S, $f^{\dagger} : A \longrightarrow X$ satisfies $f^{\dagger} = \mathcal{F}(\Delta_A); (id_A \otimes f^{\dagger}); f$.

We note that the first condition is equivalent to saying that $(-)^{\dagger}$ satisfies $(f; g)^{\dagger} = (g; f)^{\dagger}; g : I \to X$ for $f : X \to Y$ and $g : Y \to X$ in $S//A$ (page 45). Since $\mathcal{F}//A : C//A \to S//A$ has the structure in the assumption (by Lemma 6.2.3, also it is routine to see that $\mathcal{F}//A$ has a right adjoint $U//A$ given by $U//A(f) = \theta_X; U(f)$), we have just to show this simpler equation in the relativized setting, and this helps us to simplify the proof significantly.

Observe that an easier construction (c.f. Theorem 7.1.1) $Tr^X(f; \mathcal{F}(\Delta_X)) : A \longrightarrow X$ from $f : A \otimes X \longrightarrow X$ in S does not work as a fixed point operator – the construction in Theorem 7.2.1 uses the adjunction in a crucial manner.

It is in general impossible to recover a trace operator from a fixed point operator which satisfies the conditions of Theorem 7.2.1; for instance, if S has a zero object

0 such that $0 \otimes A \simeq 0$ (e.g. **Rel** below), the zero map satisfies these conditions. It is an interesting question to ask if we can strengthen the conditions so that we can recover a trace operator, though such conditions seem very delicate. In particular, in this generalized setting, our fixed point operator may not satisfy the Bekič property nor the diagonal property (see Example 7.2.8 below).

A careful inspection of our construction reveals that we need the trace operator just on objects of the form UX (equivalently $\mathcal{F}(UX)$ as $\mathcal{F}$ is identity-on-objects); actually it is sufficient if the full subcategory of S whose objects are of the form of $UX_1 \otimes \ldots \otimes UX_n$ is traced. Thus such a fixed point operator exists even in a weaker setting. It would be interesting to see if this fixed point operator determines this sub-trace structure. It would be more interesting to see if there is a good connection between such a fixed point operator and fixed point operators in models of intuitionistic linear logic as studied in [24].

An observation corresponding to Proposition 7.1.4 is

Proposition 7.2.2 In the setting as described above, assume that $h : X \to Y$ in S satisfies the following condition: for any f and g in S, if

$$
\begin{array}{ccc}
A \otimes UX & \xrightarrow{\;f\;} & B \otimes UX \\
{\scriptstyle A \otimes \mathcal{F}(Uh)} \big\downarrow & & \big\downarrow {\scriptstyle B \otimes \mathcal{F}(Uh)} \\
A \otimes UY & \xrightarrow[\;g\;]{} & B \otimes UY
\end{array}
$$

commutes, then $Tr^{UX}(f) = Tr^{UY}(g)$. Then, for any f and g, if

$$
\begin{array}{ccc}
A \otimes X & \xrightarrow{\;f\;} & X \\
{\scriptstyle A \otimes h} \big\downarrow & & \big\downarrow {\scriptstyle h} \\
A \otimes Y & \xrightarrow[\;g\;]{} & Y
\end{array}
$$

commutes then $f^{\dagger}; h = g^{\dagger}$.

Proof: See Appendix.   □

Note that our setting is equivalent to saying that we have a cartesian category C with a monad $T = U \circ \mathcal{F}$ on it, which has a commutative tensorial strength θ, such that the Kleisli category $S = C_T$ is traced. So, as we have already noted, it is possible to say that we are dealing with some *notions of computation* in the sense of Moggi [71, 72] with extra structure (trace).

Recall that higher-order acyclic sharing theories have been modeled in terms of cartesian centrally closed SMC's, whereas cyclic sharing theories were modeled by cartesian center traced SMC's. To model higher-order cyclic sharing theories, which will be presented and analyzed in the following sections, we use the combination – cartesian centrally closed traced SMC's. That is,

Definition 7.2.3 (cartesian centrally closed traced SMC)
A *cartesian centrally closed traced symmetric monoidal category* (cartesian centrally closed traced SMC) is an identity-on-objects strict symmetric monoidal functor $\mathcal{F}:$ $C \longrightarrow S$ where C is a cartesian category and S a traced symmetric monoidal category, such that for every object X the functor $\mathcal{F}(-) \otimes X : C \longrightarrow S$ has a right adjoint $X \Rightarrow$ $(-): S \longrightarrow C$. □

Remark 7.2.4 In [38] a cartesian centrally closed traced SMC is called a *traced computational model* (following Moggi's "computational models"). □

Given a cartesian centrally closed traced SMC $\mathcal{F} : C \to S$, $\mathcal{F}$ itself has a right adjoint $I \Rightarrow (-)$. Therefore

Proposition 7.2.5 In a cartesian centrally closed traced SMC $\mathcal{F} : C \to S$, there is a fixed point operator as described in Theorem 7.2.1. □

As in the case of cartesian closed categories, we also have an internalized version of the fixed point operator in cartesian centrally closed traced SMC's:

Corollary 7.2.6 Given a cartesian centrally closed traced SMC $\mathcal{F} : C \to S$, there is a dinatural transformation $\mathtt{fix}_- : \mathcal{F}((-) \Rightarrow (-)) \longrightarrow (-)$ whose components lie in S and satisfy $\mathtt{fix}_X = \mathcal{F}(\Delta_{X \Rightarrow X}); (id_{X \Rightarrow X} \otimes \mathtt{fix}_X); \mathtt{ap}_{X,X} : X \Rightarrow X \to X$.

Proof: Let $\mathtt{fix}_X$ be $\mathtt{ap}_{X,X}^{\dagger}$ and observe that $\mathcal{F}(f^*); \mathtt{fix}_X = f^{\dagger}$ for $f : A \otimes X \to X$ in S. □

To help with the intuition, we shall give a selection of cartesian centrally closed traced SMC's below. Most of them have already been mentioned in the introduction.

Example 7.2.7 (traced cartesian closed categories)
A traced cartesian closed category is a cartesian centrally closed traced SMC in which the cartesian category part and the traced category part are identical. Examples include many domain-theoretic categories such as Example 7.1.2. □

Example 7.2.8 (non-deterministic model)
The inclusion from the category **Set** of sets and functions to the category **Rel** of sets and binary relations (with the direct product of sets as the symmetric monoidal product) forms a cartesian centrally closed traced SMC: $\mathbf{Rel}(A \otimes X, B) \simeq \mathbf{Set}(A, \mathbf{Rel}(X, B))$. The trace operator on **Rel**, induced by the compact closed structure of **Rel**, is given as follows: for a relation $R : A \otimes X \longrightarrow B \otimes X$, we define a relation $Tr^X(R) : A \longrightarrow B$ by $(a, b) \in Tr^X(R)$ iff $((a, x), (b, x)) \in R$ for an $x \in X$ (here a relation from A to B is given as a subobject of $A \times B$). The parametrized fixed point operator $(-)^{\dagger}$ on **Rel** is given by

$$R^{\dagger} = \{(a, x) \mid \exists S \subseteq X \; S = \{y \mid \exists z \in S \; ((a, z), y) \in R\} \; \& \; x \in S\} : A \longrightarrow X$$

for $R : A \otimes X \longrightarrow X$ (and $R^{\dagger}$ is not the zero map!). Its internal version is given by

$$\mathtt{fix}_X = \{(R, x) \mid \exists S \subseteq X \; S; R = S \; \& \; x \in S\} : \mathbf{Rel}(X, X) \longrightarrow X.$$

This fixed point operator does *not* satisfy the diagonal property (hence Bekič property). For instance, consider $R : \text{bool} \otimes \text{bool} \to \text{bool}$ (where $\text{bool} = \{t, f\}$) such that $(t, f) \, R \, f$ and $(f, t) \, R \, t$. Then $R^\dagger : \text{bool} \to \text{bool}$ is determined by $t \, R^\dagger \, f$ and $f \, R^\dagger \, t$, therefore $R^{\dagger\dagger} \subseteq \text{bool}$ by $\{t, f\}$. However $(\Delta_{\text{bool}}; R)^\dagger = \emptyset$.   □

Note that we can use an elementary topos instead of **Set**, which may provide a computationally more sophisticated model.

An example with more classical flavour:

Example 7.2.9 (finite dimensional vector spaces over a finite field)
Let F_2 be the field with just two elements (thus its characteristic is 2), and $\textbf{Vect}_{F_2}^{\text{fin}}$ be the category of finite dimensional vector spaces (with chosen bases) over F_2. There is an identity-on-objects strict symmetric monoidal functor from the category of finite sets to $\textbf{Vect}_{F_2}^{\text{fin}}$ which maps a set S to a vector space with the basis S, and this functor has a right adjoint (the underlying functor). Since $\textbf{Vect}_{F_2}^{\text{fin}}$ is traced (in the very classical sense), this is an instance of a cartesian centrally closed traced SMC. Note that this example is similar to the previous one – compare the matrix representation of binary relations and that of linear maps.   □

Instead of 2 we can take any other prime number p and use the field F_p of p elements. If the base field is not finite, the identity-on-objects strict functor cannot have a right adjoint, thus fails to be centrally closed.

Example 7.2.10 (higher-order reflexive action calculi)
In Chapter 8 we observe that the higher-order reflexive extension of an action calculus [68, 66, 67] forms a cartesian centrally closed traced SMC. In this calculus the fixed point operator $(-)^\dagger$ is given by

$$a^\dagger = \uparrow_{\varepsilon \Rightarrow n} ((x^{\varepsilon \Rightarrow n})^\ulcorner (\text{id}_m \otimes \langle x \rangle) \cdot \textbf{ap}_{\varepsilon,n}) \cdot a^\urcorner \cdot \textbf{copy}_{\varepsilon \Rightarrow n}) \cdot \textbf{ap}_{\varepsilon,n} : m \longrightarrow n$$

for $a : m \otimes n \longrightarrow n$. Mifsud gives essentially the same operator $\text{ITER}(a)$ in his thesis [61]. Using this, we can present recursion operators in various process calculi, typically the replication operator. This issue will be further investigated in Chapter 8 (Example 8.3.9).   □

7.3   Higher-Order Cyclic Sharing Theory

Now it is almost routine to introduce higher-order cyclic sharing theories – they are obtained by combining cyclic sharing theories and higher-order ones.

Definition 7.3.1 (raw expressions)

$$M ::= x \mid F(M) \mid 0 \mid M_1 \otimes M_2 \mid \text{letrec } (x_1, \ldots, x_m) \text{ be } M_1 \text{ in } M_2 \mid \lambda x.M \mid M_1 M_2$$

□

Definition 7.3.2 (values)

$$V ::= 0 \mid x \mid \lambda(\vec{x}).M \mid V_1 \otimes V_2$$

$\square$

Definition 7.3.3 (typing)

$$\frac{}{\Gamma, x:\sigma \vdash x:(\sigma)} \text{ variable}$$

$$\frac{\Gamma \vdash M:(\sigma_1,\ldots,\sigma_m) \quad F:(\sigma_1,\ldots,\sigma_m) \to (\tau_1,\ldots,\tau_n)}{\Gamma \vdash F(M):(\tau_1,\ldots,\tau_n)} \text{ operator}$$

$$\frac{}{\Gamma \vdash 0:()} \text{ unit}$$

$$\frac{\Gamma \vdash M:(\sigma_1,\ldots,\sigma_m) \quad \Gamma \vdash N:(\tau_1,\ldots,\tau_n)}{\Gamma \vdash M \otimes N:(\sigma_1,\ldots,\sigma_m,\tau_1,\ldots,\tau_n)} \text{ tensor}$$

$$\frac{\begin{array}{c}\Gamma, x_1:\sigma_1,\ldots,x_m:\sigma_m \vdash M:(\sigma_1,\ldots,\sigma_m) \\ \Gamma, x_1:\sigma_1,\ldots,x_m:\sigma_m \vdash N:(\tau_1,\ldots,\tau_n)\end{array}}{\Gamma \vdash \text{letrec } (x_1,\ldots,x_m) \text{ be } M \text{ in } N:(\tau_1,\ldots,\tau_n)} \text{ letrec}$$

$$\frac{\Gamma, x:\sigma, x':\sigma', \Gamma' \vdash M:(\tau_1,\ldots,\tau_n)}{\Gamma, x':\sigma', x:\sigma, \Gamma' \vdash M:(\tau_1,\ldots,\tau_n)} \text{ exchange}$$

$$\frac{\Gamma, \vec{x}:\vec{\sigma} \vdash M:(\vec{\tau})}{\Gamma \vdash \lambda(\vec{x}).M:((\vec{\sigma}) \Rightarrow (\vec{\tau}))} \text{ abstraction}$$

$$\frac{\Gamma \vdash M:((\vec{\sigma}) \Rightarrow (\vec{\tau})) \quad \Gamma \vdash N:(\vec{\sigma})}{\Gamma \vdash MN:(\vec{\tau})} \text{ application}$$

$\square$

Definition 7.3.4 (axioms)

(id)	letrec $(\vec{x})$ be M in $\vec{x}$	$-$	$M \quad (\vec{x} \notin FV(M))$
(ass$_1$)	letrec $(\vec{x})$ be (letrec $(\vec{y})$ be L in M) in N	$=$	letrec $(\vec{x},\vec{y})$ be $M \otimes L$ in N
(ass$_2$)	letrec $(\vec{x})$ be L in letrec $(\vec{y})$ be M in N	$=$	letrec $(\vec{x},\vec{y})$ be $L \otimes M$ in N
($\otimes_1$)	$L \otimes$ (letrec $(\vec{x})$ be M in N)	$=$	letrec $(\vec{x})$ be M in $L \otimes N$
($\otimes_2$)	(letrec $(\vec{x})$ be L in M) $\otimes N$	$=$	letrec $(\vec{x})$ be L in $M \otimes N$
(perm)	letrec $(\vec{x},\vec{y},\vec{z})$ be $M_1 \otimes M_2 \otimes M_3$ in N	$=$	letrec $(\vec{y},\vec{x},\vec{z})$ be $M_2 \otimes M_1 \otimes M_3$ in N
(subst)	letrec $(\vec{x})$ be M in $F(N)$	$=$	$F(\text{letrec } (\vec{x}) \text{ be } M \text{ in } N)$
(β)	$(\lambda(\vec{x}).M)N$	$=$	letrec $(\vec{x})$ be N in M
(η_0)	$\lambda(\vec{x}).y(\vec{x})$	$=$	y
(app$_1$)	(letrec $(\vec{x})$ be L in M)N	$=$	letrec $(\vec{x})$ be L in MN
(app$_2$)	$L(\text{letrec } (\vec{x}) \text{ be } M \text{ in } N)$	$=$	letrec $(\vec{x})$ be M in LN
(deref)	letrec $(x,\vec{y})$ be $V \otimes M$ in N	$=$	letrec $(x,\vec{y})$ be $V \otimes (M\{V/x\})$ in $N\{V/x\}$
(g.c.)	letrec (x) be V in N	$=$	$N \quad (x \notin FV(N) \cup FV(V))$

Both sides of axioms must have the same type under the same context. $\square$

Definition 7.3.5 (higher-order cyclic sharing theory)
A *higher-order cyclic sharing theory* over Σ is an equational theory on the well-typed terms closed under the term constructions described above, where the equality on terms is a congruence relation containing the axioms above. By the *pure higher-order cyclic sharing theory*, we mean the higher-order cyclic sharing theory with no additional axioms.   □

As in the first-order case, for readability, we introduce the following syntax for the multiple letrec-binding:

$$\frac{\Gamma,\vec{x}_1 : \vec{\sigma}_1,\ldots,\vec{x}_k : \vec{\sigma}_k \vdash M_i : (\vec{\sigma}_i) \quad (1 \leq i \leq k) \qquad \Gamma,\vec{x}_1 : \vec{\sigma}_1,\ldots,\vec{x}_k : \vec{\sigma}_k \vdash N : (\vec{\tau})}{\Gamma \vdash \text{letrec } (\vec{x}_1) \text{ be } M_1,\ldots (\vec{x}_k) \text{ be } M_k \text{ in } N : (\vec{\tau})}$$

for letrec $(\vec{x}_1,\ldots,\vec{x}_k)$ be $M_1 \otimes \ldots \otimes M_k$ in N.

Remark 7.3.6 All axioms except those for substitution of values are inherited from the axioms of cyclic sharing theories and higher-order acyclic sharing theories, which form the first two groups respectively. The axioms for substitution of values, which we name (deref) and (g.c.), come from those for cyclic sharing theories (page 33), and this time we allow the substitutions of values as the axiom (σ_v) of the higher-order acyclic theory. We note that our axioms for substitutions are as strong as "one-step replacement" style substitutions, see below.   □

Lemma 7.3.7 The below are derivable:

$$\text{letrec } (x,\vec{y}) \text{ be } V \otimes M[x] \text{ in } N \;=\; \text{letrec } (x,\vec{y}) \text{ be } V \otimes M[V] \text{ in } N$$
$$\text{letrec } (x,\vec{y}) \text{ be } V \otimes M \text{ in } N[x] \;=\; \text{letrec } (x,\vec{y}) \text{ be } V \otimes M \text{ in } N[V]$$

where $[x]$ indicates an occurrence of free x in the expression.

Proof: We demonstrate the first case.

$$
\begin{aligned}
&\quad \text{letrec } (x) \text{ be } V, (\vec{y}) \text{ be } M[x] \text{ in } N \\
&= \text{letrec } (x) \text{ be } (\text{letrec } (x') \text{ be } V \text{ in } x'), (\vec{y}) \text{ be } M[x] \text{ in } N && \text{(id)} \\
&= \text{letrec } (x) \text{ be } x', (x') \text{ be } V, (\vec{y}) \text{ be } M[x'] \text{ in } N && \text{(ass}_1) \\
&= \text{letrec } (x') \text{ be } V, (x) \text{ be } x', (\vec{y}) \text{ be } M[x'] \text{ in } N && \text{(perm)} \\
&= \text{letrec } (x') \text{ be } V, (x) \text{ be } V, (\vec{y}) \text{ be } M[V] \text{ in } N && \text{(deref)} \\
&= \text{letrec } (x) \text{ be } V, (\vec{y}) \text{ be } M[V] \text{ in } N && \text{(g.c.)}
\end{aligned}
$$

□

By the same proofs as in Chapter 4 and 6, we have

Lemma 7.3.8 α-conversions of letrec-bindings and lambda-bindings, and β and η axioms restricted to values are derivable.   □

The notions of structures and models are defined in the same manner as in Chapter 3, 4 and 6.

Theorem 7.3.9 (soundness)
Let $[\![-]\!]$ be a higher-order cyclic sharing model of a higher-order cyclic sharing theory. If $\Gamma \vdash M = N : (\vec{\sigma})$ is derivable in the theory, then $[\![\Gamma \vdash M : (\vec{\sigma})]\!] = [\![\Gamma \vdash N : (\vec{\sigma})]\!]$. $\square$

Also we have the classifying category $\mathcal{F}_\mathbb{T} : C_\mathbb{T} \to S_\mathbb{T}$ for a higher-order cyclic sharing theory $\mathbb{T}$ and a complete model (generic model) in it.

Theorem 7.3.10 (completeness)
Given a theory $\mathbb{T}$, there is a complete model in $\mathcal{F}_\mathbb{T} : C_\mathbb{T} \to S_\mathbb{T}$, given by $[\![\sigma]\!] = \sigma$ and $[\![F]\!] = [\vec{x} : \vec{\sigma} \vdash F(\vec{x}) : (\vec{\tau})]$ for $F : (\vec{\sigma}) \to (\vec{\tau})$. $\square$

Example 7.3.11 The fixed point operator in Theorem 7.2.1 can be represented using the higher-order cyclic sharing theory as an internal language of such categorical structures, as

$$\frac{\Gamma, \vec{x} : \vec{\sigma} \vdash M : (\vec{\sigma})}{\Gamma \vdash \mu(\vec{x}).M \ \equiv \ \text{letrec } (f) \text{ be } \lambda().(\text{letrec } (\vec{x}) \text{ be } f0 \text{ in } M) \text{ in } f0 : (\vec{\sigma})}$$

Let us show dinaturality using the equational theory. Assume that $\Gamma, \vec{x} : \vec{\sigma} \vdash M : (\vec{\tau})$ and $\Gamma, \vec{y} : \vec{\tau} \vdash N : (\vec{\sigma})$. Then we have compositions of them $M \circ N$ and $N \circ M$ by

$$\Gamma, \vec{x} : \vec{\sigma} \vdash M \circ N \ \equiv \ \text{letrec } (\vec{y}) \text{ be } M \text{ in } N : (\vec{\sigma})$$

and

$$\Gamma, \vec{y} : \vec{\tau} \vdash N \circ M \ \equiv \ \text{letrec } (\vec{x}) \text{ be } N \text{ in } M : (\vec{\tau}).$$

Also we define values V_M and V_N by

$$\Gamma, f : () \Rightarrow (\vec{\sigma}) \vdash V_M \ \equiv \ \lambda().(\text{letrec } (\vec{x}) \text{ be } f0 \text{ in } M) : (() \Rightarrow (\vec{\tau}))$$

and

$$\Gamma, g : () \Rightarrow (\vec{\tau}) \vdash V_N \ \equiv \ \lambda().(\text{letrec } (\vec{y}) \text{ be } g0 \text{ in } N) : (() \Rightarrow (\vec{\sigma})).$$

Then we can show

$$\Gamma, f : () \Rightarrow (\vec{\sigma}) \vdash \text{letrec } (g) \text{ be } V_M \text{ in } V_N \ = \ \lambda().(\text{letrec } (\vec{x}) \text{ be } f0 \text{ in } N \circ M) : (() \Rightarrow (\vec{\sigma}))$$

$$\Gamma, g : () \Rightarrow (\vec{\tau}) \vdash \text{letrec } (f) \text{ be } V_N \text{ in } V_M \ = \ \lambda().(\text{letrec } (\vec{y}) \text{ be } g0 \text{ in } M \circ N) : (() \Rightarrow (\vec{\tau}))$$

using β and axioms for substitutions. Now dinaturality is proved as

$$
\begin{aligned}
&\mu(\vec{x}).(N \circ M) \\
=\ &\text{letrec } (f) \text{ be } \lambda().(\text{letrec } (\vec{x}) \text{ be } f0 \text{ in } N \circ M) \text{ in } f0 \\
=\ &\text{letrec } (f) \text{ be } (\text{letrec } (g) \text{ be } V_M \text{ in } V_N) \text{ in } f0 \\
=\ &\text{letrec } (f) \text{ be } V_N, (g) \text{ be } V_M \text{ in } f0 && (\text{ass}_1) \\
=\ &\text{letrec } (f) \text{ be } V_N, (g) \text{ be } V_M \text{ in } V_N 0 && (\text{deref}) \\
=\ &\text{letrec } (f) \text{ be } V_N, (g) \text{ be } V_M \text{ in } \text{letrec } (\vec{y}) \text{ be } g0 \text{ in } N && (\beta_v) \\
=\ &\text{letrec } (\vec{y}) \text{ be } (\text{letrec } (g) \text{ be } V_M, (f) \text{ be } V_N \text{ in } g0) \text{ in } N && (\text{ass}_1) \\
=\ &\text{letrec } (\vec{y}) \text{ be } (\text{letrec } (g) \text{ be } (\text{letrec } (f) \text{ be } V_N \text{ in } V_M) \text{ in } g0) \text{ in } N \\
=\ &\text{letrec } (\vec{y}) \text{ be } (\text{letrec } (g) \text{ be } \lambda().(\text{letrec } (\vec{y}) \text{ be } g0 \text{ in } M \circ N) \text{ in } g0) \text{ in } N \\
=\ &\text{letrec } (\vec{y}) \text{ be } \mu(\vec{y}).(M \circ N) \text{ in } N.
\end{aligned}
$$

$\square$

7.4   Cyclic Lambda Calculi

As a fragment and a variant of higher-order cyclic sharing theories, we introduce two simply typed lambda calculi enriched with the notion of cyclic sharing, the *simply typed* λ_{letrec}*-calculus* and λ^v_{letrec}*-calculus* in which cyclically shared resources are represented in terms of the letrec syntax. Traced cartesian closed categories and cartesian centrally traced SMCs serve as sound and complete models of these calculi respectively.

Basically we deal with a simplified pure higher-order cyclic sharing theory with no multiple conclusion and no operator symbols, thus almost all observations here can be seen as consequences of the results obtained so far. However, since cyclic lambda calculi have their own right to be studied, especially in the connection with graph rewriting theory and implementations of functional programming languages, we spell out them here.

The Syntax and Axioms

We design the simply typed λ^v_{letrec}-calculus as a modification of (the commutative version of) Moggi's computational lambda calculus (Definition 5.3.4) [71]; we replace the let-syntax by the letrec-syntax which allows cyclic bindings.

In this section, we fix a set of *base types*.

Types

$$\sigma,\tau\ldots\ ::=\ b\mid\sigma\Rightarrow\tau\quad\text{(where } b \text{ is a base type)}$$

Syntactic Domains

Variables	$x,y,z\ldots$
Raw Terms	$M,N\ldots\ ::=\ x\mid\lambda x.M\mid MN\mid\text{letrec } D \text{ in } N$
Values	$V,W\ldots\ ::=\ x\mid\lambda x.M$
Declarations	$D\ldots\ ::=\ x=M\mid x=M,D$

In a declaration, binding variables are assumed to be disjoint.

Typing

$$\frac{}{\Gamma,x:\sigma\vdash x:\sigma}\ \text{Variable}\qquad\frac{\Gamma,x:\sigma,y:\sigma',\Gamma'\vdash M:\tau}{\Gamma,y:\sigma',x:\sigma,\Gamma'\vdash M:\tau}\ \text{Exchange}$$

$$\frac{\Gamma,x:\sigma\vdash M:\tau}{\Gamma\vdash\lambda x.M:\sigma\Rightarrow\tau}\ \text{Abstraction}\qquad\frac{\Gamma\vdash M:\sigma\Rightarrow\tau\quad\Gamma\vdash N:\sigma}{\Gamma\vdash MN:\tau}\ \text{Application}$$

$$\frac{\Gamma,x_1:\sigma_1,\ldots,x_n:\sigma_n\vdash M_i:\sigma_i\ (i=1,\ldots,n)\quad\Gamma,x_1:\sigma_1,\ldots,x_n:\sigma_n\vdash N:\tau}{\Gamma\vdash\text{letrec } x_1=M_1,\ldots,x_n=M_n \text{ in } N:\tau}\ \text{letrec}$$

Axioms

Identity	letrec $x = M$ in x	$=$	M $(x \notin FV(M))$
Associativity	letrec $y = ($letrec D_1 in $M), D_2$ in N $=$		letrec $D_1, y = M, D_2$ in N
	letrec D_1 in letrec D_2 in M	$=$	letrec D_1, D_2 in M
Permutation	letrec D_1, D_2, D in N	$=$	letrec D_2, D_1, D in N
Commutativity	(letrec D in $M)N$	$=$	letrec D in MN
	$M($letrec D in $N)$	$=$	letrec D in MN
β	$(\lambda x.M)N$	$=$	letrec $x = N$ in M
σ_v	letrec $x = V, D[x]$ in M	$=$	letrec $x = V, D[V]$ in M
	letrec $x = V, D$ in $M[x]$	$=$	letrec $x = V, D$ in $M[V]$
	letrec $x = V$ in M	$=$	M $(x \notin FV(V) \cup FV(M))$
η_0	$\lambda x.yx$	$=$	y

Both sides of equations must have the same type under the same typing context; we will work just on well-typed terms. We assume the usual conventions on variables.

We remark that axioms Identity, Associativity, Permutation and Commutativity ensure that two $\lambda^v_{\text{letrec}}$-terms are identified if they correspond to the same cyclic directed graph; thus they are a sort of structural congruence, rather than representing actual computation. β creates a sharing from a function application. σ_v describes the substitution of values (the first two for the dereference, the last one for the garbage collection). In $M[x]$ and $D[x]$, $[x]$ denotes a free occurrence of x. From β, σ_v and η_0, we have the "call-by-value" $\beta\eta$-equations:

Lemma 7.4.1 In $\lambda^v_{\text{letrec}}$-calculus, the following are derivable.

$$
\begin{array}{llll}
\beta_v & (\lambda x.M)V & = & M\{V/x\} \\
\eta_v & (\lambda x.Vx) & = & V \quad (x \notin FV(V)) \quad \square
\end{array}
$$

We think it is misleading to relate this calculus to the call-by-value operational semantics; restricting substitutions on values does not mean that this calculus is call-by-value. Rather, our equational theory is fairly close to the *call-by-need* calculus proposed in [9], which corresponds to a version of lazy implementations of the call-by-name operational semantics. We expect that this connection is the right direction to relate our calculus to an operational semantics.

Also we define a "strengthened" version in which arbitrary substitution and η-reduction are allowed (thus any term is a value):

σ	letrec $x = N, D[x]$ in M	$=$	letrec $x = N, D[N]$ in M
	letrec $x = N, D$ in $M[x]$	$=$	letrec $x = N, D$ in $M[N]$
	letrec $x = N$ in M	$=$	M $(x \notin FV(M))$
η	$\lambda x.Mx$	$=$	M $(x \notin FV(M))$

We shall call this version the *simply typed* λ_{letrec}-*calculus* – this corresponds to the calculus in [10] ignoring the typing and the extensionality (η-axiom).

Interpretation into Models

We spell out how to interpret our cyclic lambda calculi in traced cartesian closed categories as well as cartesian centrally closed traced SMC's. Again it is essentially a simplification of the previous interpretations of sharing theories.

We just present the case of the $\lambda^v_{\text{letrec}}$-calculus; the case of the λ_{letrec}-calculus is obtained just by replacing a cartesian centrally closed traced SMC by a traced cartesian closed category.

Let us fix a cartesian centrally closed traced SMC $\mathcal{F} : C \longrightarrow S$, and choose an object $[\![b]\!]$ for each base type b. The interpretation of arrow types is then defined by $[\![\sigma \Rightarrow \tau]\!] = [\![\sigma]\!] \Rightarrow [\![\tau]\!]$. We interpret a $\lambda^v_{\text{letrec}}$-term (with its typing environment) $x_1 : \sigma_1, \ldots, x_n : \sigma_n \vdash M : \tau$ by an arrow $[\![x_1 : \sigma_1, \ldots, x_n : \sigma_n \vdash M : \tau]\!] : [\![\sigma_1]\!] \otimes \ldots \otimes [\![\sigma_n]\!] \longrightarrow [\![\tau]\!]$ in S as follows.

$$[\![x_1 : \sigma_1, \ldots, x_n : \sigma_n \vdash x_i : \sigma_i]\!] = \mathcal{F}\pi_i \text{ where } \pi_i \text{ is the } i\text{-th projection}$$
$$[\![\Gamma \vdash \lambda x.M : \sigma \Rightarrow \tau]\!] = \mathcal{F}(([\![\Gamma, x : \sigma \vdash M : \tau]\!])^*)$$
$$[\![\Gamma \vdash M^{\sigma \Rightarrow \tau} N^\sigma : \tau]\!] = \mathcal{F}\Delta; ([\![\Gamma \vdash M : \sigma \Rightarrow \tau]\!] \otimes [\![\Gamma \vdash N : \tau]\!]); \mathbf{ap}$$
$$[\![\Gamma \vdash \text{letrec } x_1 = M_1^{\sigma_1}, \ldots, x_k = M_k^{\sigma_k} \text{ in } N : \tau]\!] =$$
$$\mathcal{F}\Delta; (id \otimes Tr^{[\![\sigma_1]\!] \otimes \ldots [\![\sigma_k]\!]}(\mathcal{F}\Delta_k; ([\![\Gamma' \vdash M_1 : \sigma_1]\!] \otimes \ldots [\![\Gamma' \vdash M_k : \sigma_k]\!]); \mathcal{F}\Delta)); [\![\Gamma' \vdash N : \tau]\!]$$

Recall that $\mathbf{ap}_{A,B} : (A \Rightarrow B) \otimes A \longrightarrow B$ is the counit of the adjoint $\mathcal{F}(-) \otimes A \dashv A \Rightarrow (-)$, and $(-)^* : \mathcal{T}(\mathcal{F}A \otimes B, C) \longrightarrow C(A, B \Rightarrow C)$ is the associated natural bijection. In the last case, Γ' is $\Gamma, x_1 : \sigma_1, \ldots, x_k : \sigma_k$ and Δ_{kA} is the k-times copy from A to $\underbrace{A \times \ldots \times A}_{k \text{ times}}$. Note that values are first interpreted in C (following Moggi's account, C is the category of values) and then lifted to S via $\mathcal{F}$.

Calculations similar to those in previous chapters show that cartesian centrally closed traced SMCs are sound for the $\lambda^v_{\text{letrec}}$-calculus (and the same for traced cartesian closed categories and the λ_{letrec}-calculus):

Theorem 7.4.2 (Soundness)

- For any cartesian centrally closed traced SMC with chosen object $[\![b]\!]$ for each base type b, this interpretation is sound; if $\Gamma \vdash M : \sigma$, $\Gamma \vdash N : \sigma$ and $M = N$ in the $\lambda^v_{\text{letrec}}$-calculus then $[\![\Gamma \vdash M : \sigma]\!] = [\![\Gamma \vdash N : \sigma]\!]$.

- For any traced cartesian closed category with chosen object $[\![b]\!]$ for each base type b, this interpretation is sound; if $\Gamma \vdash M : \sigma$, $\Gamma \vdash N : \sigma$ and $M = N$ in the λ_{letrec}-calculus then $[\![\Gamma \vdash M : \sigma]\!] = [\![\Gamma \vdash N : \sigma]\!]$. □

Example 7.4.3 (domain-theoretic model)

As we already noted, **Dom** is a traced cartesian closed category (hence also a cartesian centrally closed traced SMC). The interpretation of a λ_{letrec}-term $\vdash$ letrec $x = M$ in $x : \sigma$ in **Dom** is just the least fixed point $\bigcup_n F^n(\bot)$ where $F : [\![\sigma]\!] \longrightarrow [\![\sigma]\!]$ is the interpretation of $x : \sigma \vdash M : \sigma$. □

Similarly, the traced cartesian closed category of ω-cpo's with bottoms serves as a sound model of the λ_{letrec}-calculus. Since there is a faithful interpretation of the simply typed lambda calculus (with no constant) in this category (due to Plotkin, c.f. Theorem 2 of [83]) which factors through the interpretation of the λ_{letrec}-calculus, we have

Corollary 7.4.4 The λ_{letrec}-calculus is a conservative extension of the simply typed lambda calculus. □

which can be stated more semantically as

Proposition 7.4.5 The free cartesian closed category generated by a set of objects faithfully embeds into the free traced cartesian closed category (with the same objects). □

Remark 7.4.6 For non-free cases, adding a trace while preserving the cartesian (closed) structure may cause a collapse. Plotkin (private communication) has shown that there is no finite product preserving embedding from a cartesian category with a Mal'cev operator into a traced cartesian category, by appealing to the observation by Plotkin and Simpson (Theorem 10 in [79]) that algebraic theories with a Mal'cev operator and a fixed point operator are inconsistent. □

Example 7.4.7 (non-deterministic model)
In **Rel** (Example 7.2.8), a $\lambda^{\nu}_{\text{letrec}}$-term is interpreted as the set of "all possible solutions of the recursive equation". The interpretation of $\vdash$ letrec $x = M$ in $x : \sigma$ is just the set $\{x \in [\![\sigma]\!] \mid (x,x) \in [\![x : \sigma \vdash M : \sigma]\!]\}$ (a subobject of $[\![\sigma]\!] = 1 \times [\![\sigma]\!]$). For instance,

$$
\begin{aligned}
[\![\vdash \text{letrec } x = x \text{ in } x : \sigma]\!] &= [\![\sigma]\!] &&: 1 \longrightarrow [\![\sigma]\!] \\
[\![\vdash \text{letrec } x = x^2 \text{ in } x : \text{nat}]\!] &= \{0,1\} &&: 1 \longrightarrow \mathbf{N} \\
[\![\vdash \text{letrec } x = x+1 \text{ in } x : \text{nat}]\!] &= \emptyset &&: 1 \longrightarrow \mathbf{N}
\end{aligned}
$$

(for the latter two cases we enrich the calculus with natural numbers). Note that this model is sound for the $\lambda^{\nu}_{\text{letrec}}$-calculus, but not for the λ_{letrec}-calculus – since we cannot copy non-deterministic computation, this model is "resource sensitive". □

This non-deterministic model is not a complete (faithful) model of the $\lambda^{\nu}_{\text{letrec}}$-calculus. However, we conjecture that the commutative computational lambda calculus faithfully embeds into this model, which implies that the $\lambda^{\nu}_{\text{letrec}}$-calculus is conservative over the commutative computational lambda calculus.

The classifying category of the pure higher-order sharing theory serves as a complete model of the $\lambda^{\nu}_{\text{letrec}}$-calculus. Thus we also have completeness:

Theorem 7.4.8 (Completeness)

- If $[\![\Gamma \vdash M : \sigma]\!] = [\![\Gamma \vdash N : \sigma]\!]$ for every cartesian centrally closed traced SMC, then $M = N$ in the $\lambda^{\nu}_{\text{letrec}}$-calculus.

- If $[\![\Gamma \vdash M : \sigma]\!] = [\![\Gamma \vdash N : \sigma]\!]$ for every traced cartesian closed category, then $M = N$ in the λ_{letrec}-calculus. □

Remark 7.4.9 To represent the parametrized fixed point operator given in Theorem 7.2.1 we have to extend the $\lambda^{\nu}_{\text{letrec}}$-calculus with a *unit type* unit which has a unique value $*$:

$$\frac{}{\Gamma \vdash * : \text{unit}} \text{ Unit} \qquad V = * \quad (V : \text{unit})$$

The interpretation of the unit type in a cartesian centrally closed traced SMC is just the terminal object (unit object). The type constructor unit $\Rightarrow (-)$ then plays the role of the right adjoint of the inclusion from the category of values to the category of terms. We define the parametrized fixed point operator by

$$\frac{\Gamma, x : \sigma \vdash M : \sigma}{\Gamma \vdash \mu x^{\sigma}.M \equiv \text{letrec } f^{\text{unit} \Rightarrow \sigma} = \lambda y^{\text{unit}}.((\lambda x^{\sigma}.M)(f*)) \text{ in } f* \ : \ \sigma}$$

which satisfies $\mu x.M = (\lambda x.M)(\mu x.M)$, but may not satisfy the standard fixed point equation $\mu x.M = M\{\mu x.M/x\}$ in the $\lambda^{\nu}_{\text{letrec}}$-calculus because $\mu x.M$ may not be a value in general. The operator Y_3 in the next section is essentially the same as this fixed point operator, except for avoiding to use unit. $\qquad\qquad\square$

We could give the untyped version and its semantic models – by a reflexive object in a cartesian centrally closed traced SMC (or a traced cartesian closed category). Regarding the results in previous sections, we can establish the connection between the dinatural diagonal fixed point operator in a model of the untyped λ_{letrec}-calculus and the trace operator of the cartesian closed category. It would be interesting to compare recursion created by untypedness and recursion created by trace (cyclic sharing) in such models.

Comparison with Ariola-Blom-Klop Approach

A detailed study of the equational and rewriting-theoretic aspects of cyclic lambda calculi has been done by Ariola and Blom [8], following the previous work by Ariola and Klop [10]. Our cyclic lambda calculi fit into their account, but a few remarks should follow. Our equational theories (for the λ_{letrec}- and $\lambda^{\nu}_{\text{letrec}}$-calculi) are strictly weaker than the corresponding equational theories in [8] ignoring the eta axiom η_0. This is because

- Our equational theories correspond to what Ariola and Blom call the "scoped lambda graphs" where two lambda graphs are separated if they have different scoping of variable bindings even when they have the same underlying graph. Ariola and Blom introduce axioms for identifying them.

- Ariola and Blom introduce more axioms to equate graphs which are bisimular, i.e. have the same (infinite) tree unwinding.

- For the $\lambda^{\nu}_{\text{letrec}}$-calculus, the notion of values is slightly different, in that Ariola and Blom treat cyclic terms which contain only values in a similar manner to values.

Assuming η_0, their equational theory is a quotient of ours. Ariola and Blom have shown that their calculus is sound and complete for their infinite tree unwinding semantics (this is a highly non-trivial result as syntactically cyclic lambda calculi have no confluent rewriting system [10], therefore no easy term model exists; Ariola and Blom overcome this problem by showing a semantic confluence result up to the contents of "information" in terms). We conjecture that their semantic models give rise to a traced cartesian closed category and a cartesian centrally closed traced SMC.

Therefore, equationally our (pure) theories are essentially proper subsets of the "most complete" calculi in [8]. However, interestingly, as rewriting systems, ours are as strong as theirs; Ariola and Blom have shown in [8] that, regarded as rewriting systems, our calculi are complete for their infinite semantics. It is interesting to see if the infinite semantics, conjectured to form instances of our models, stand out as models with some "good" characterization – at least we should like to know the theoretical justification of why infinite trees models have occupied an important position as the canonical semantic models of rewriting systems for long time.

7.5 Analyzing Fixed Points

In the $\lambda^{\nu}_{\text{letrec}}$-calculus, several (weak) fixed point operators are definable – this is not surprising, because there are several known encodings of fixed point operators in terms of cyclic sharing. However, it is difficult to see that they are not identified by our equational theory – syntactic reasoning for cyclic graph structures is not an easy task, as the non-confluency result in [10] suggests. On the other hand, in many traditional models for recursive computation, all of them have the same denotational meaning mainly because we cannot distinguish values from non-values in such models.

One purpose in developing the models of higher-order cyclic sharing theories is to give a clear semantic account for these several recursive computations created from cyclic sharing. Though this topic has not yet been fully developed, we shall give some elementary analysis using the $\lambda^{\nu}_{\text{letrec}}$-calculus and a model (**Rel**).

We define $\lambda^{\nu}_{\text{letrec}}$-terms $\Gamma \vdash Y_i(M) : \sigma$ $(i = 1,2,3)$ for given term $\Gamma \vdash M : \sigma \Rightarrow \sigma$ as follows.

$$
\begin{array}{lcl}
Y_1 & = & \text{letrec fix}^{(\sigma\Rightarrow\sigma)\Rightarrow\sigma} = \lambda f^{\sigma\Rightarrow\sigma}.f(\text{fix } f) \text{ in fix} \\
Y_2 & = & \lambda f^{\sigma\Rightarrow\sigma}.\text{letrec } x^{\sigma} = fx \text{ in } x \\
Y_3(M) & = & \text{letrec } g^{\tau\Rightarrow\sigma} = \lambda y^{\tau}.M(gy) \text{ in } gN \\
& & (N \text{ is a closed term of type } \tau, \text{ e.g. letrec } x = x \text{ in } x : \tau)
\end{array}
$$

Each of them can be used as a fixed point operator, but their behaviours are not the same. For instance, it is known that Y_2 is more efficient than others, under the call-by-need evaluation strategy [57]. Y_1 satisfies the fixed point equation $YV = V(YV)$ for any value $V : \sigma \Rightarrow \sigma$.

$$
\begin{array}{lcll}
Y_1 M & = & \text{letrec fix} = \lambda f.f(\text{fix } f) \text{ in fix} M & \text{Commutativity} \\
& = & \text{letrec fix} = \lambda f.f(\text{fix } f) \text{ in } (\lambda f.f(\text{fix } f))M & \sigma_{\nu} \\
& = & \text{letrec fix} = \lambda f.f(\text{fix } f) \text{ in letrec } f' = M \text{ in } f'(\text{fix } f') & \beta \\
& = & \text{letrec } f' = M \text{ in letrec fix} = \lambda f.f(\text{fix } f) \text{ in } f'(\text{fix } f') & \text{Assoc., Perm.} \\
& = & \text{letrec } f' = M \text{ in } f'((\text{letrec fix} = \lambda f.f(\text{fix } f) \text{ in fix})f') & \text{Commutativity} \\
& = & \text{letrec } f' = M \text{ in } f'(Y_1 f') & \\
(& = & M(Y_1 M) \quad \text{if } M \text{ is a value}) &
\end{array}
$$

Y_2 satisfies $Y_2M = M(Y_2M)$ only when Mx is equal to a value (hence M is supposed to be a higher-order value). If $M = \lambda y.V$ for some value V,

$$
\begin{aligned}
Y_2M &= \text{letrec } x = (\lambda y.V)x \text{ in } x & \beta_v \\
&= \text{letrec } x = V\{x/y\} \text{ in } x & \beta_v \\
&= \text{letrec } x = V\{x/y\} \text{ in } V\{x/y\} & \sigma_v \\
&= \text{letrec } x = (\lambda y.V)x \text{ in } (\lambda y.V)x & \beta_v \\
&= (\lambda y.V)(\text{letrec } x = (\lambda y.V)x \text{ in } x) & \text{Commutativity} \\
&= M(Y_2M)
\end{aligned}
$$

Y_3 satisfies $Y_3(M) = M(Y_3(M))$ for any term $M : \sigma \Rightarrow \sigma$ (thus is a "true" fixed point operator).

$$
\begin{aligned}
Y_3(M) &= \text{letrec } g = \lambda y.M(gy) \text{ in } gN \\
&= \text{letrec } g = \lambda y.M(gy) \text{ in } (\lambda y.M(gy))N & \sigma_v \\
&= \text{letrec } g = \lambda y.M(gy) \text{ in letrec } y' = N \text{ in } M(gy') & \beta \\
&= \text{letrec } g = \lambda y.M(gy) \text{ in } M(g(\text{letrec } y' = N \text{ in } y')) & \text{Commutativity} \\
&= M(\text{letrec } g = \lambda y.M(gy) \text{ in } g(\text{letrec } y' = N \text{ in } y')) & \text{Commutativity} \\
&= M(\text{letrec } g = \lambda y.M(gy) \text{ in } gN) & \text{Identity} \\
&= M(Y_3(M))
\end{aligned}
$$

The interpretation of these operators in a cartesian centrally closed traced SMC is as follows.

$$
\begin{aligned}
[\![\vdash Y_1]\!] &= Tr^{(A \Rightarrow A) \mapsto A}(F(\mathbf{cur}((id \otimes \Delta); (\mathbf{ap} \otimes id); c; \mathbf{ap})); \Delta) \\
[\![\vdash Y_2]\!] &= F(\mathbf{cur}(Tr^A(\mathbf{ap}; \Delta))) \\
[\![\Gamma \vdash Y_3(M)]\!] &= (Tr^{B \mapsto A}(F(\mathbf{cur}(([\![\Gamma \vdash M : \sigma \Rightarrow \sigma]\!] \otimes \mathbf{ap}); \mathbf{ap})); \Delta) \otimes [\![\vdash N : \tau]\!]); \mathbf{ap}
\end{aligned}
$$

where $A = [\![\sigma]\!]$ and $B = [\![\tau]\!]$. They have the different interpretations in **Rel**, hence are not identified in the $\lambda_{\text{letrec}}^v$-calculus. Assume that $S = [\![\vdash M : \sigma \Rightarrow \sigma]\!] \subseteq \mathbf{Rel}(A, A)$. Then

$$
[\![\vdash Y_1(M) : \sigma]\!] = \bigcup_{f \in S} \bigcup_{(A'; f) = A' \subseteq A} A' \qquad [\![\vdash Y_2(M) : \sigma]\!] = \bigcup_{f \in S} \{x \mid (x, x) \in f\}
$$

whereas

$$
[\![\vdash Y_3(M) : \sigma]\!] = \bigcup_{(A'; \bigcup S) = A' \subseteq A} A'
$$

(In the definition of Y_3, we take $N : \tau$ as letrec $x = x$ in $x : \tau$.)

Remark 7.5.1 If we interpret a λ_{letrec}-term of the form letrec $x = M$ in N by an untyped term $(\lambda x.N)(Y(\lambda x.M))$ where $Y = \lambda f.(\lambda x.f(xx))(\lambda x.f(xx))$ (Curry's fixed point combinator), the fixed point operators above are related to more or less familiar combinators (c.f. [14], Chapter 6.1):

- Y_1 corresponds to $(\lambda x.\lambda y.y(xxy))(\lambda x.\lambda y.y(xxy))$ which is Turing's fixed point combinator.

- Y_2 corresponds to $\lambda f.(\lambda x.f(xx))(\lambda x.f(xx))$ which is Curry's fixed point combinator.

- $\lambda f.Y_3(f)$ corresponds to $\lambda f.(\lambda x.\lambda y.f(xxy))(\lambda x.\lambda y.f(xxy))N$.

$\square$

8

Action Calculi

We show that our framework for sharing graphs can accommodate Milner's *action calculi* [68], a proposed framework for general interactive computation, by showing that our sharing theories, enriched with suitable constructs for interpreting parameterized constants called controls, are equivalent to the action calculi and their higher-order/reflexive extensions [66, 67, 61].

The dynamics, the computational counterpart of action calculi, is then understood as rewriting systems on our theories, and interpreted as local preorders on our models. In this sense, we understand action calculi as generalized graph rewriting systems – and regard the notion of sharing as one of the fundamental concepts of action calculi.

We first review the definition of action calculi. We then extend our sharing theories to accommodate action calculi. The essential point is to introduce the parametrized operator symbols which correspond to controls. The semantic observations immediately tell us that we are dealing with the equivalent beings and there are obvious syntactic translations between sharing theories and action calculi. This observation extends to the higher-order and reflexive (cyclic) extensions of action calculi equally well.

8.1 Action Calculi: Definitions, Basics

We introduce an action calculus as a quotient of a term algebra. An action calculus is defined by a set of typed terms, an equational theory on it (theory AC) and a preorder on the equivalence classes of terms, called a *reaction relation* or *dynamics*.

We first fix a freely generated monoid $M = (M, \otimes, \varepsilon)$; M's elements will be called *arities* and generators will be called *prime arities*. We also assume a set X; its elements are called *names*, and we assign a prime arity to each name. If an arity p is assigned to a name x, we write $x : p$. Further, we assume that infinitely many names are associated with each prime arity.

An action calculus $\mathsf{AC}(\mathcal{K})$ is then specified by a set $\mathcal{K}$ of *controls* (*control operators*) each of which is equipped with an *arity rule* and a reaction relation which is usually generated by a few *reaction rules*.

Definition 8.1.1 The set of *terms* over $\mathcal{K}$ is generated by the following rules.
[Raw terms]

$$a, b \ldots \quad ::= \quad \langle x \rangle \mid (x)a \mid \mathbf{id} \mid a \cdot b \mid a \otimes b \mid K(a_1, \ldots, a_k)$$

[Arity assignment (Typing)]

$$\frac{x:p}{\langle x\rangle : \varepsilon \to p} \quad \frac{a:m\to n \quad x:p}{(x)a : p\otimes m\to n} \quad \frac{}{\mathbf{id}_m : m\to m} \quad \frac{a:k\to l \quad b:l\to m}{a\cdot b:k\to m}$$

$$\frac{a:k\to m \quad b:l\to n}{a\otimes b:k\otimes l\to m\otimes n} \quad \frac{a_i:m_i\to n_i \quad (i=1,\ldots,k)}{K(a_1,\ldots,a_k):m\to n}$$

where each control $K\in\mathcal{K}$ is equipped with arity $((m_1,n_1),\ldots,(m_k,n_k),(m,n))$ which may be subject to some side conditions; there can be some dependency between the number of arguments (k) and the arities m_i, n_i, m and n.   □

We may omit the arity subscripts of terms if there is no confusion. The notions of *free* and *bound name* are defined as usual; a name x is bound in $(x)a$, and $\langle x\rangle$ is considered a free occurrence of x. The set of free names in a term a will be denoted by $fn(a)$.

Definition 8.1.2 We use the following derived notations for multiple names and a notation for derived terms (permutation).

$$(x_1,\ldots,x_n)a \triangleq (x_1)\ldots(x_n)a$$
$$\langle x_1,\ldots,x_n\rangle \triangleq \langle x_1\rangle\otimes\ldots\otimes\langle x_n\rangle$$

$$\mathbf{p}_{m,n} \triangleq (\vec{x},\vec{y})\langle\vec{y},\vec{x}\rangle : m\otimes n\to n\otimes m$$

In $(x_1,\ldots,x_n)a$, names x_i are assumed to be all different.   □

Definition 8.1.3 The equational theory AC is the set of equations upon terms generated by the following axioms.

[The strict monoidal category axioms]

A1	$a\cdot\mathbf{id}=a=\mathbf{id}\cdot a$	A4	$a\cdot(b\cdot c)=(a\cdot b)\cdot c$
A2	$a\otimes\mathbf{id}_\varepsilon=a=\mathbf{id}_\varepsilon\otimes a$	A5	$a\otimes(b\otimes c)=(a\otimes b)\otimes c$
A3	$\mathbf{id}\otimes\mathbf{id}=\mathbf{id}$	A6	$(a\cdot b)\otimes(a'\cdot b')=(a\otimes a')\cdot(b\otimes b')$

[The concrete axioms]

σ $(\langle y\rangle\otimes\mathbf{id}_m)\cdot(x)a=a\{y/x\}$ $(a:m\to n)$

δ $(x)(((\langle x\rangle\otimes\mathbf{id}_m)\cdot a)=a$ $(x:p,a:p\otimes m\to n, x\notin fn(a))$

ζ $\mathbf{p}_{k,m}\cdot(b\otimes a)=(a\otimes b)\cdot\mathbf{p}_{l,n}$ $(a:k\to l,b:m\to n)$

We call the equivalence classes of terms *actions*, for which we overload the same notations for terms and actions provided there is no ambiguity; a may represent a term or the equivalence class of the term depending on the context.   □

We will write $\mathsf{AC}(\mathcal{K})(m,n)$ for the set of actions of arity $m\to n$.

Definition 8.1.4 The *action calculus* $\mathsf{AC}(\mathcal{K})$ is the equational theory given as above together with a preorder $\searrow_{m,n}$ (subscripts may be omitted) on $\mathsf{AC}(\mathcal{K})(m,n)$ for each m and n, called *reaction relation* or *dynamics*, which is closed under tensor, composition and abstraction such that identities are minimal, i.e. $\mathbf{id}\searrow a$ implies $a=\mathbf{id}$.   □

Note that controls may not preserve the reaction relation. The definition of theory AC presented here is different from the original version in [68] in the choice of primitives and axioms, but it is easy to verify that they give the same equational theory. Axioms A1~A6 imply that $AC(\mathcal{K})$ is a strict monoidal category whose objects are arities and arrows are actions. The concrete axioms σ and δ determine how names and abstractions work, whereas from ζ we can show that $AC(\mathcal{K})$ is a symmetric monoidal category with symmetry **p**.

The definition of dynamics implies that

Proposition 8.1.5 If $a \searrow b$ and $a \neq b$ then a must contain a control. $\qquad\qquad \Box$

Therefore any reaction must involve some controls.

Before proceeding to the technical discussion, let us briefly recall how a simple polyadic π-calculus [64] (originally introduced by Tokoro and Honda as the ν-calculus [43], or the asynchronous π-calculus) can be represented as an action calculus. For details, see [68].

Example 8.1.6 (the action calculus PIC)
The action calculus $AC(\textbf{box}, \textbf{out})$, or PIC, is determined by the following data.

First, we assume that arities are freely generated from just one element - that is, we consider the monoid of the additive structure on natural numbers. Thus we have just one prime arity 1, and assume $\varepsilon = 0$, $m \otimes n = m + n$. Then we introduce controls for input and output bindings:

$$\textbf{out} : 1 + m \to 0 \qquad \frac{a : m \to n}{\textbf{box}(a) : 1 \to n}$$

with the reaction relation generated by a reaction rule

$$(\langle x \rangle \cdot \textbf{box}(a)) \otimes (((\langle x \rangle \otimes \textbf{id}) \cdot \textbf{out}) \quad \searrow \quad a.$$

$\qquad\qquad \Box$

In [68] Milner has given a translation from a simple π-calculus into PIC which preserves the structural congruence and dynamics.

8.2   Action Calculi as Sharing Theories

We show that an action calculus is equivalent to a pure acyclic sharing theory enriched with parametrized operator symbols which correspond to control operators in action calculi. This is archived by observing that these two syntactic theories share the same semantic models.

Parametrized Operators

First we revise our definition of signature (Definition 2.1.1) so that we can accommodate parametrized operator symbols which correspond to control operators in action calculi.

Definition 8.2.1 (signature)

Let S be a set of sorts. An *(extended) S-sorted signature* is a set Σ of operation symbols together with an arity function assigning to each operation symbol K a list

$$((\vec{\sigma}_1,\vec{\tau}_1),\dots,(\vec{\sigma}_r,\vec{\tau}_r),(\vec{\sigma},\vec{\tau}))$$

of pairs of finite lists of elements of S. $\square$

If $r = 0$, we recover the original definition of signatures of (non-parametrized) operators. Recall that the typing rule for an operator symbol $F : (\vec{\sigma}) \to (\vec{\tau})$ is given as

$$\frac{\Gamma \vdash M : (\vec{\sigma})}{\Gamma \vdash F(M) : (\vec{\tau})} \ \text{operator}$$

Together with the parameters we extend this as

$$\frac{\Gamma,\vec{x}_i : \vec{\sigma}_i \vdash M_i : (\vec{\tau}_i) \ (i \le r) \quad \Gamma \vdash M : (\vec{\sigma})}{\Gamma \vdash K((\vec{x}_1)M_1,\dots,(\vec{x}_r)M_r \mid M) : (\vec{\tau})} \ \text{operator}$$

where K is an operator with arity $((\vec{\sigma}_1,\vec{\tau}_1),\dots,(\vec{\sigma}_r,\vec{\tau}_r),(\vec{\sigma},\vec{\tau}))$. The axioms (Definition 2.2.3) make sense for parametrized operators (with an obvious modification on (subst)), but we need an additional axiom (α_{op}) for α-conversions of locally bound variables:

(subst)	let $(\vec{x})$ be M in $K(\dots \mid N)$	$=$	$K(\dots \mid$ let $(\vec{x})$ be M in $N)$
(α_{op})	$K(\dots,(\vec{x}_i)M_i,\dots \mid N)$	$=$	$K(\dots,(\vec{y}_i)M_i\{\vec{y}_i/\vec{x}_i\},\dots \mid N)$
			($\vec{y}$ are fresh variables)

Both sides of axioms must have the same type under the same context.

Let us add some comments on parametrized operators. A parametrized operator K has two different sorts of arguments - one for arguments with some local quantification of names $((\vec{x}_i)s_i)$ and the other for a non-quantified argument (t). The most popular example of the former is lambda-abstraction: we can write the usual abstraction rule as

$$\frac{\Gamma,x : \sigma \vdash M : \tau}{\Gamma \vdash \lambda((x)M \mid 0) : \sigma \Rightarrow \tau}$$

instead of $\lambda x.M$. On the other hand, an application of a function symbol does not need any locally quantified arguments, like the following.

$$\frac{\Gamma \vdash M : \mathsf{nat}}{\Gamma \vdash \mathbf{succ}(\mid M) : \mathsf{nat}}$$

Another example is the input-binding in the (polyadic) π-calculus (recall the action calculus AC(**box**,**out**) in the last section)

$$\frac{\Gamma,\vec{y} : m \vdash P : n \quad \Gamma \vdash x : p}{\Gamma \vdash \mathbf{box}((\vec{y})P \mid x) : n}$$

which is usually written as $x(\vec{y}).P$. The elimination rule for disjunction is also a good instance:

$$\frac{\Gamma, x : \sigma_1 \vdash L : \tau \quad \Gamma, y : \sigma_2 \vdash M : \tau \quad \Gamma \vdash N : \sigma_1 \vee \sigma_2}{\Gamma \vdash \mathbf{case}((x)L, (y)M \mid N) : \tau}$$

Therefore parametrized operators can be understood as term constructors which may involve the local bindings of variables. Gordon Plotkin points out that such operators are analogues of Aczel's general binding operators [6].

As we noted, α-conversion on let-bound variables can be derived from other axioms. However, we need an axiom for α-conversion of locally bound names in controls ($\mathcal{K}_\alpha$), which cannot be derived from the other axioms.

Interpretation into Models

We give the semantic interpretation of parametrized operators (equivalently control constants) as natural families of functions (or natural transformations) on a cartesian-center SMC, c.f. [34, 75]. We extend the definition of structure (Definition 3.2.9) as follows.

For each operator symbol K with its arity rule $((m_1, n_1), \ldots, (m_k, n_k), (m, n))$, we assume a family of functions

$$[\![K]\!]_X : S(\mathcal{F}(X) \otimes [\![m_1]\!], [\![n_1]\!]) \times \ldots \times S(\mathcal{F}(X) \otimes [\![m_k]\!], [\![n_k]\!]) \longrightarrow S(\mathcal{F}(X) \otimes [\![m]\!], [\![n]\!])$$

natural in the parameter X in C.

Using this extended definition, we can define the semantic interpretation of the parametrized operators:

$$[\![\Gamma \vdash K((\vec{x}_1)M_1, \ldots \mid N) : (\vec{\tau})]\!] = \begin{array}{l} \mathcal{F}\Delta; (id \otimes [\![\Gamma \vdash N : (\vec{\tau})]\!]); \\ [\![K]\!]_{[\![\Gamma]\!]}([\![\Gamma, \vec{x}_1 : \vec{\sigma}_1 \vdash M_1 : (\vec{\tau}_1)]\!], \ldots) \end{array}$$

The following result follows from [75]:

Theorem 8.2.2 *The sharing models of a pure acyclic sharing theory gives rise to models of the action calculus with the same signature.* □

Corollary 8.2.3 *There are sound and complete translations between a pure acyclic sharing theory and the corresponding action calculus, which are inverse to each other.* □

The detail of syntactic translations are found in [35].

Remark 8.2.4 The following examples show the difference between the traditional function symbols (non-parametrized operator symbols as given in Chapter 2) and our parametrized operator symbols. A function symbol F with its arity rule $((m, n))$ is equipped with a term construction

$$\frac{\Gamma \vdash M : m}{\Gamma \vdash F(\mid M) : n}$$

which is interpreted as an arrow $[\![F]\!] : [\![m]\!] \to [\![n]\!]$ in S of a cartesian-center SMC $\mathcal{F}$: $C \to S$. On the other hand, a parametrized operator G with its arity rule $((0,m),(0,n))$ has a corresponding term construction

$$\frac{\Gamma \vdash M : m}{\Gamma \vdash G(()M \mid 0) : n}$$

which is interpreted as a family of functions $[\![G]\!]_X : S(\mathcal{F}X, [\![m]\!]) \to S(\mathcal{F}X, [\![n]\!])$ natural in X in C. If we assume the naturality in S then (by Yoneda) there is a bijection between these two interpretations, thus there is no significance in having a parametrized operator G. However, as the naturality is restricted to C, $[\![G]\!]$'s interpretation is not determined by $[\![G]\!]_{[\![m]\!]}(id)$, thus has an extra generality. In the equational theory, $[\![G]\!]$ behaves like a function symbol which does not satisfy the axiom (subst). After constructing the classifying category in Chapter 3 we observed that the axiom (subst) has nothing to do with the completeness; without (subst), we get an interpretation of G rather than F, which possibly contains more junk arrows arising from the extra freedom of G over F.   ☐

The dynamics of an action calculus has semantic interpretation in a cartesian-center SMC as a local preorder on it [75].

Theorem 8.2.5 The following are equivalent:

- dynamics on an action calculus,

- rewriting systems with the minimality condition on the corresponding sharing theory (Chapter 2) satisfying the "strengthening"

$$\frac{\Gamma \vdash M : (\vec{\tau}) \quad \Gamma \vdash N : (\vec{\tau}) \quad \Gamma, x : \sigma \vdash M \succ N : (\vec{\tau})}{\Gamma \vdash M \succ N : (\vec{\tau})}$$

   and

- local preorders with the minimality condition on the classifying category (Chapter 3) satisfying that $id_X \otimes f \rightsquigarrow id_X \otimes g$ implies $f \rightsquigarrow g$ for any X.[1]

                                                                ☐

8.3 Extensions

We recall the higher-order and reflexive extensions of the action calculi, and extend the correspondence between action calculi and sharing theories to these settings.

Higher-Order Action Calculi

In [66] Milner extends action calculi to *higher-order action calculi*, which are intended to be a unified framework for higher-order concurrent computation. There have been

---

[1]This condition is pointed out by Philippa Gardner (private communication).

many proposed applications of the calculi to various concurrent/functional models, including higher-order extensions of process calculi, petri nets (proposed in [66]) and call-by-value/call-by-name variants of PCF [47].

In this section we observe that higher-order action calculi (with a mild refinement) correspond precisely to our (pure) higher-order acyclic sharing theory because they share the same semantic models. Some of these results are reported also in [35].

Example 8.3.1 More complicated arities are needed for representing a higher-order extension of the action calculus, which will be described below. For such a purpose, the arities will be generated from a set of base arities $(b, \ldots)$ as follows.

$$\begin{aligned} \text{prime arities} \quad p, q \ldots \quad &::= \quad b \mid m \Rightarrow n \\ \text{arities} \quad m, n \ldots \quad &::= \quad \varepsilon \mid p \mid m \otimes n \end{aligned}$$

Intuitively, arities correspond to the types of the simply typed lambda calculus with (strictly associative) products. The arrow types are included in the prime arities because we need names of arrow types for representing higher-order communication.

Then we introduce two new controls. Given a set of controls $\mathcal{K}$ we define $\mathcal{K}^{\Rightarrow} \triangleq \mathcal{K} \cup \{ \ulcorner - \urcorner, \mathbf{ap} \}$ where the controls $\ulcorner - \urcorner$ and $\mathbf{ap}$ are subject to the arity rules below.

$$\frac{a : m \to n}{\ulcorner a \urcorner : \varepsilon \to m \Rightarrow n} \qquad \frac{}{\mathbf{ap} : (m \Rightarrow n) \otimes m \to n}$$

Now we can define the dynamics for higher-order computation on this extended action calculus $\mathrm{AC}(\mathcal{K}^{\Rightarrow})$ by the following reaction rules.

$$\begin{aligned} \beta \qquad (\ulcorner a \urcorner \otimes \mathbf{id}) \cdot \mathbf{ap} \quad &\searrow \quad a \\ \sigma_{\mathrm{code}} \qquad (\ulcorner a \urcorner \otimes \mathbf{id}) \cdot (x)b \quad &\searrow \quad b\{\ulcorner a \urcorner / x\} \end{aligned}$$

Controls are assumed to preserve β and σ_{code} reactions. In this calculus, we can pass not only a name but also a *code* $\ulcorner a \urcorner$ of an action a as a higher-order datum (σ_{code} reaction). The code can be applied or decoded with help of $\mathbf{ap}$ (β reaction). □

Following [66], we proceed to define the higher-order action calculus. Since our concern is on the equational characterization of such a general framework of higher-order communication, we naturally regard the reaction rules β and σ as axioms rather than abstract rewriting rules – we refer [66] for some rewriting theoretical aspects of these reaction rules.

Definition 8.3.2 (higher-order action calculus)
The *higher-order action calculus* $\mathrm{HAC}(\mathcal{K})$ is obtained as the quotient of the action calculus $\mathrm{AC}(\mathcal{K}^{\Rightarrow})$ (see the last example) by axioms β, σ_{code} and η_{name} below

$$\begin{aligned} \beta \qquad (\ulcorner a \urcorner \otimes \mathbf{id}) \cdot \mathbf{ap} \quad &= \quad a \\ \sigma_{\mathrm{code}} \qquad (\ulcorner a \urcorner \otimes \mathbf{id}) \cdot (x)b \quad &= \quad b\{\ulcorner a \urcorner / x\} \\ \eta_{\mathrm{name}} \qquad \ulcorner (\langle x \rangle \otimes \mathbf{id}) \cdot \mathbf{ap} \urcorner \quad &= \quad \langle x \rangle \end{aligned}$$

which is equipped with (the $\beta \sigma_{\mathrm{code}}$-quotient of) the dynamics of $\mathrm{AC}(\mathcal{K}^{\Rightarrow})$. □

There is an axiom which is not included in the original version [66]: the axiom η_{name}. At first sight, the reader may think this is rather an artificial change. However, for codes we can derive the η-equation

$$\ulcorner(\ulcorner a\urcorner \otimes \mathbf{id}) \cdot \mathbf{ap}\urcorner = \ulcorner a\urcorner$$

from the β-axiom. Since codes and names are treated in the same way (as "values") in the higher-order action calculi, and since this rather weak η-axiom does not change the syntactic nature of the calculi (the conservativity from the first-order calculi still holds; see below), we decide to include it in our version. Actually this is a needed change to get clearer semantic models – with this axiom higher-order action calculi allow a good category-theoretical characterization. Note that a stronger η-axiom

$$\eta_{\text{too-strong}} \qquad \ulcorner(a \otimes \mathbf{id}) \cdot \mathbf{ap}\urcorner \ = \ a$$

makes the calculus cartesian closed [66], thus changes the equational theory significantly. This is because $\eta_{\text{too-strong}}$ equates a value (the code of the left hand side) to a (possibly) non-value (right hand side). Since the models we are interested in may not be cartesian closed, we do not accept this axiom in general. For further discussion, see [66]. Anyway we have a better observation:

Theorem 8.3.3 The semantic models of a higher-order action calculus are given by those of the pure higher-order sharing theory with the same signature in cartesian centrally closed SMC's.   □

Detail is found in [35]. This implies the syntactic equivalence (translations) between pure higher-order sharing theories and higher-order action calculi.

We already know that the action calculus embeds into the higher-order calculus by comparing molecular forms [66]. The semantic proof of this result is available, as we have for the pure acyclic sharing theory and pure higher-order acyclic sharing theory, by constructing a model of the higher-order theory from that of the first-order one (Chapter 5). We need a little care with parametrized operators (controls): for simplicity, we just consider the one-parameter case. Assume that $\mathcal{F} : C \to S$ is a cartesian-center SMC, and $[\![-]\!]$ is a model in it, and we have an operator K of arity rule $((m_1, n_1), (m, n))$. Then there is a family of functions $[\![K]\!]_X : S(\mathcal{F}(X) \otimes [\![m_1]\!], [\![n_1]\!]) \longrightarrow S(\mathcal{F}(X) \otimes [\![m]\!], [\![n]\!])$ natural in X in C. Following Corollary 5.2.2, we have a cartesian centrally closed SMC $\bar{\mathcal{F}} : \bar{C} \to \bar{S}$ with a fully faithful cartesian-center functor (i_C, i_S) from $\mathcal{F} : C \to S$ to $\bar{\mathcal{F}} : \bar{C} \to \bar{S}$, where i_C is dense. Since i_S is fully faithful strict symmetric monoidal and $\bar{\mathcal{F}} \circ i_C = i_S \circ \mathcal{F}$, this induces a family of functions $[\![K]\!]'_X : \bar{S}(\bar{\mathcal{F}}(i_C(X)) \otimes i_S([\![m_1]\!]), i_S([\![n_1]\!])) \longrightarrow \bar{S}(\bar{\mathcal{F}}(i_C(X)) \otimes i_S([\![m]\!]), i_S([\![n]\!]))$ natural in X in C. Since i_C is dense, we can extend $[\![K]\!]'$ to a family of functions $[\![K]\!]'_X : \bar{S}(\bar{\mathcal{F}}(X) \otimes i_S([\![m_1]\!]), i_S([\![n_1]\!])) \longrightarrow \bar{S}(\bar{\mathcal{F}}(X) \otimes i_S([\![m]\!]), i_S([\![n]\!]))$ natural in X in $\bar{C}$. Following the same discussion in Chapter 5, we obtain the proof of conservativity.

Reflexive Action Calculi

In [67] Milner extends action calculi to allow cyclic bindings of names by introducing an operator called *reflexion*. It may be of some surprise that his notion of reflexion turns out to be equivalent to traces of symmetric monoidal categories. This observation, by Alex Mifsud and myself, first reported in Mifsud's thesis [61], enables us to see that (cartesian center) traced monoidal categories serve as models of reflexive action calculi, and also to accommodate reflexive action calculi as cyclic sharing theories.

Definition 8.3.4 (reflexive action calculus)
The *reflexive action calculus* $\mathsf{AC}^r(\mathcal{K})$ [67] is obtained as the quotient of the action calculus $\mathsf{AC}(\uparrow, \mathcal{K})$ by axioms $\rho_1 \sim \rho_6$ as below.

1. The control $\uparrow$ is given by

$$\frac{a : p \otimes m \to p \otimes n}{\uparrow_p^{m,n}(a) : m \to n}$$

 (in the sequel, subscripts may be omitted). Unlike the usual controls, $\uparrow$ is assumed to preserve the reaction relation.

2. The reflexion axioms:

$$
\begin{array}{lll}
\rho_1 & \mathbf{id}_p & = & \uparrow_p(\mathbf{p}_{p,p}) \\
\rho_2 & \uparrow_p(a) \otimes \mathbf{id} & = & \uparrow_p(a \otimes \mathbf{id}) \\
\rho_3 & \uparrow_p(a) \cdot b & = & \uparrow_p(a \cdot (\mathbf{id}_p \otimes b)) \\
\rho_4 & a \cdot \uparrow_p(b) & = & \uparrow_p((\mathbf{id}_p \otimes a) \cdot b) \\
\rho_6 & \uparrow_q(\uparrow_p(a)) & = & \uparrow_p(\uparrow_q((\mathbf{p}_{q,p} \otimes \mathbf{id}) \cdot a \cdot (\mathbf{p}_{p,q} \otimes \mathbf{id}))) \\
\end{array}
$$

Remark 8.3.5 We drop the axiom ρ_5 included in the original definition, which is derivable from other axioms (c.f. [61]). $\qquad\square$

In the definition above, the reflexion operator is defined only on prime arities. We extend it to general arities by

$$
\begin{array}{lll}
\uparrow_\varepsilon^{m,n}(a) & \equiv & a \\
\uparrow_{p \otimes l}^{m,n}(a) & \equiv & \uparrow_l^{m,n}(\uparrow_p^{l \otimes m, l \otimes n}(a)).
\end{array}
$$

For ease of explanation, we extend the definition of reflexion to a general strict symmetric monoidal category.

Definition 8.3.6 Let C be a strict symmetric monoidal category and suppose that $Obj(C)$ is freely generated from a set of objects $\mathcal{P}$. A *reflexion* on C is a family of functions

$$\uparrow_p^{A,B} : C(P \otimes A, P \otimes B) \longrightarrow C(A,B)$$

where A, B are objects of C and P varies in $\mathcal{P}$, subject to the axioms $\rho_1 \sim \rho_6$. $\qquad\square$

A simple and pleasant fact is that traces and reflexions are precisely the same:

Proposition 8.3.7 Let C be a strict symmetric monoidal category whose objects are freely generated from a set of objects $\mathcal{P}$. Then C is traced if and only if it has a reflexion.   □

The proof is elementary but lengthy – see [61] for the full calculation.

Corollary 8.3.8 A reflexive action calculus is traced.   □

This, together with the observations on action calculi and acyclic sharing theories, immediately implies that the models of a reflexive action calculus are given by models of the corresponding pure cyclic sharing theory in cartesian-center traced SMC's. Therefore we have the correspondence between reflexive action calculi and pure cyclic sharing theories.

Combining the results above, we also have the correspondence between a higher-order reflexive action calculus and the pure higher-order cyclic sharing theory with the same signature, which share the same models in cartesian centrally closed traced SMC's.

The combination of higher-order and reflexive features enables us to encode recursive programs – this is essentially a consequence of results in Chapter 7. We give an example to motivate some intuition, as well as to make a connection with previous work by Mifsud [61].

Example 8.3.9 In functional languages with a fixed-point operator Y, we can define the infinite application (loop) of a function M as $Y(\lambda x.\lambda y.M(xy))$ or $Y(\lambda x.\lambda y.x(My))$. If the language allows us to use letrec-bindings (as in cyclic sharing theories), they can be written as letrec $x = \lambda y.M(xy)$ in x and letrec $x = \lambda y.x(My)$ in x respectively. We shall demonstrate these simple forms of iteration exist in the higher-order reflexive action calculus; this is a variation of the observations in the last chapter.

Let $a : m \longrightarrow m$ be an action in the higher-order reflexive action calculus. In [61] Mifsud defines actions $\text{ITER}(a)$ and $\text{BACKITER}(a)$ by

$$\frac{a : m \longrightarrow m \quad x : n \Rightarrow m \quad x \notin fn(a)}{\text{ITER}(a) \equiv (\text{rec}((x)^{\ulcorner}(\langle x \rangle \otimes \mathbf{id}) \cdot \mathbf{ap} \cdot a^{\urcorner}) \otimes \mathbf{id}) \cdot \mathbf{ap} : n \longrightarrow m}$$

$$\frac{a : m \longrightarrow m \quad x : m \Rightarrow n \quad x \notin fn(a)}{\text{BACKITER}(a) \equiv (\text{rec}((x)^{\ulcorner}(\langle x \rangle \otimes a) \cdot \mathbf{ap}^{\urcorner}) \otimes \mathbf{id}) \cdot \mathbf{ap} : m \longrightarrow n}$$

where $\text{rec}_p(a)$ is given by

$$\frac{a : p \otimes m \longrightarrow p \otimes n}{\text{rec}_p(a) \equiv \uparrow_p (a \cdot (\mathbf{copy}_p \otimes \mathbf{id}_n)) : m \longrightarrow p \otimes n}$$

They satisfy the equations below.

$$\text{ITER}(a) = \text{ITER}(a) \cdot a$$

$$\text{BACKITER}(a) = a \cdot \text{BACKITER}(a)$$

These equations are easily and intuitively verified in the higher-order cyclic sharing theories. Assume that we have a closed term $M : m \Rightarrow m$ of the higher-order cyclic sharing theory which satisfies $[\![\vdash M]\!] = \ulcorner a \urcorner$ where $[\![-]\!]$ indicates the translation from the higher-order cyclic sharing theory to the higher-order reflexive action calculus. Then we have

$$\text{ITER}(a) = [\![\vec{z} : n \vdash \text{letrec } (x) \text{ be } \lambda(\vec{y}).M(x\vec{y}) \text{ in } x\vec{z} : m]\!]$$

$$\text{BACKITER}(a) = [\![\vec{z} : m \vdash \text{letrec } (x) \text{ be } \lambda(\vec{y}).x(M\vec{y}) \text{ in } x\vec{z} : n]\!]$$

For instance, one can prove $\text{ITER}(a) = \text{ITER}(a) \cdot a$ as follows. In the sharing theory, we have

$$
\begin{aligned}
& \text{letrec } (x) \text{ be } \lambda(\vec{y}).M(x\vec{y}) \text{ in } x\vec{z} \\
= \; & \text{letrec } (x) \text{ be } \lambda(\vec{y}).M(x\vec{y}) \text{ in } (\lambda(\vec{y}).M(x\vec{y}))\vec{z} && \text{(deref)} \\
= \; & \text{letrec } (x) \text{ be } \lambda(\vec{y}).M(x\vec{y}) \text{ in } M(x\vec{z}) && \beta_v \\
= \; & M(\text{letrec } (x) \text{ be } \lambda(\vec{y}).M(x\vec{y}) \text{ in } x\vec{z}) && (\text{app}_2)
\end{aligned}
$$

Hence

$$
\begin{aligned}
& \text{ITER}(a) \\
= \; & [\![(\vec{z}) : n \vdash \text{letrec } (x) \text{ be } \lambda(\vec{y}).M(x\vec{y}) \text{ in } x\vec{z} : m]\!] \\
= \; & [\![(\vec{z}) : n \vdash M(\text{letrec } (x) \text{ be } \lambda(\vec{y}).M(x\vec{y}) \text{ in } x\vec{z}) : m]\!] \\
= \; & ([\![\vdash M : m \Rightarrow m]\!] \otimes [\![(\vec{z}) : n \vdash \text{letrec } (x) \text{ be } \lambda(\vec{y}).M(x\vec{y}) \text{ in } x\vec{z} : m]\!]) \cdot \mathbf{ap} \\
= \; & [\![(\vec{z}) : n \vdash \text{letrec } (x) \text{ be } \lambda(\vec{y}).M(x\vec{y}) \text{ in } x\vec{z} : m]\!] \cdot a \\
= \; & \text{ITER}(a) \cdot a.
\end{aligned}
$$

$\square$

ITER and BACKITER are not new in the context of functional programming languages; in [31], Filinski shows how to get recursion from BACKITER (the loop combinator) and first-class continuations in the call-by-value setting.

Remark 8.3.10 In his thesis [61], Mifsud asks whether ITER and BACKITER satisfy some universal property. We have a partial and positive answer for his question in some specific models. First, in many domain theoretic models, in which traces are given by least fixed point operators, ITER and BACKITER are the least ones satisfying the recursive equations. On the other hand, in our **Rel**-semantics, they are represented as

$$[\![\text{ITER}(a)]\!] = \bigcup_{\substack{f : [\![n]\!] \to [\![m]\!] \\ f ; [\![a]\!] = f}} f \quad \text{and} \quad [\![\text{BACKITER}(a)]\!] = \bigcup_{\substack{f : [\![m]\!] \to [\![n]\!] \\ [\![a]\!] ; f = f}} f$$

Thus, for any $g : [\![n]\!] \longrightarrow [\![m]\!]$ (resp. $[\![m]\!] \longrightarrow [\![n]\!]$) such that $g ; [\![a]\!] = g$ (resp. $[\![a]\!] ; g = g$), we have $g \subseteq [\![\text{ITER}(a)]\!]$ (resp. $g \subseteq [\![\text{BACKITER}(a)]\!]$), hence they are the universal (greatest) invariants of the relation $[\![a]\!] : m \longrightarrow m$ (w.r.t. the inclusion of relations). $\square$

9
Conclusion

In this thesis, we have developed a theory of models of sharing graphs arising from graph rewriting theory. Generalizing the traditional theory-model correspondence between algebraic theories and finite product preserving functors, we have established the connection between theories for sharing graphs and their models described in terms of symmetric monoidal categories, strict symmetric monoidal functors and additional requirements (adjunctions, and traces). As an important case study, we have looked at recursive computation modeled in our higher-order cyclic sharing theories and their models. Also we have shown that Milner's action calculi can be understood in terms of our sharing theories enriched with parametrized operators. As an interesting implication, our axiomatic treatment of the classes of models has enabled us to compare them with those for related theories, including Moggi's notions of computation as well as intuitionistic linear type theory.

One important piece of work yet to be done is to strengthen the connection with rewriting theory. While our work establishes the equational foundation of sharing graphs, the analysis of rewriting on our theories is still to be done. However, as the interplay between the study of cartesian closed categories and the study of rewriting systems on lambda terms turned out to be fruitful, we hope that future work will show that our theories and models provide some useful insights into the rewriting theories. A recent work by Benaissa, Lascanne and Rose [17] on sharing graphs is based on the idea of explicit substitutions [1]. Since the origin of explicit substitutions is closely related with categorical combinators on cartesian closed categories [28], we expect that a similar story can be derived from our categorical models.

Yet another important direction to be examined is the extension to the premonoidal setting [77]. While symmetric monoidal structure serves well as the models of sharing graphs for which we do not assume any specific ordering of computation on resources, that seems to be too loose for modeling more sophisticated models of computation in which some ordering of computation should be specified. The most interesting situation seems to be the traced (symmetric) premonoidal categories. We have a preliminary definition of this notion and have shown the results corresponding to the structural theorem in [50] and the fixed point theorem in this thesis (Theorem 7.2.1) and [38]. We expect that this generalized setting will be useful for analyzing recursive computation created from imperative features (like Landin's applicative order imperative fixed point operator, c.f. [33]), as well as the interplay between recursion and continuations [31, 86]. We have a few preliminary observations on modeling a language with ML-like states in cartesian centrally closed traced symmetric *premonoidal* categories, where the imperative encoding of recursion can be analyzed.

Apart from these investigations into models of computation, the structures dealt with in this thesis provide us with several mathematically interesting questions, especially in connection with traced monoidal categories. For example, while the conservativity results on acyclic theories are relatively easily established (Chapter 5), we know very few results for the cyclic cases. As stated in Theorem 6.1.6, Plotkin has shown that there is a symmetric monoidal category to which we cannot add a trace without causing a collapse. He also has given counterexamples for cartesian categories (Remark 7.4.6). Yet we do not have answers to lots of similar questions, some have already been mentioned in Chapter 6 and 7. The difficulty of these problems seems to be deeply related to the difficulty of dealing with recursion semantically. We do not even know any generic way to construct traced categories (or categories with fixed point operators).

The role of action calculi in this thesis is somewhat, at least for the author, delicate. Since one may regard our sharing theories as a simplified action calculi (without parametrized controls), it is possible to view this thesis as devoted to the models of action calculi. We did not take that direction because we wanted to develop a self-contained coherent theory which is relatively independent of any specific computational interpretation. Action calculi are "calculi for interaction", but our results seem to have no direct connection with the notion of "interaction". However, we acknowledge that we got many fundamental ideas from the work on action calculi by many people [68, 61, 34, 42, 75], and also hope that our results provide useful feedback to the study of action calculi.

A
Proofs

A.1 Proof of Proposition 6.1.5

(IF part) Assume $f : A \otimes U \to B \otimes U$ and $g : B \otimes U \to C \otimes U$.

$$
\begin{aligned}
& Tr^U(f); Tr^U(g) \\
=\; & Tr^U(f;(Tr^U(g) \otimes id_U)) & \text{Right Tightening} \\
=\; & Tr^U(f; Tr^U((id_B \otimes c_{U,U}); (g \otimes id_U); (id_C \otimes c_{U,U}))) & \text{Superposing} \\
=\; & Tr^U(f; Tr^U((id_B \otimes c_{U,U}); (g \otimes id_U))) & c_{U,U} = id_U \\
=\; & Tr^U(f; Tr^U(id_B \otimes c_{U,U}); g) & \text{Right Tightening} \\
=\; & Tr^U(f; (id_B \otimes Tr^U(c_{U,U})); g) & \text{Superposing} \\
=\; & Tr^U(f;g) & \text{Yanking}
\end{aligned}
$$

(ONLY IF part) First, functoriality of Tr^U implies that

$$
f = id_U \otimes Tr^U(f) \text{ for any } f : U \to U \; (*)
$$

as

$$
\begin{aligned}
& f \\
=\; & f; Tr^U(c_{U,U}) & \text{Yanking} \\
=\; & Tr^U((f \otimes id_U); c_{U,U}) & \text{Right Tightening} \\
=\; & Tr^U(c_{U,U}; (id_U \otimes f)) & \\
=\; & Tr^U(c_{U,U}); Tr^U(c_{U,U}; (id_U \otimes f)) & \text{Yanking} \\
=\; & Tr^U(c_{U,U}; c_{U,U}; (id_U \otimes f)) & \text{by the functoriality} \\
=\; & Tr^U(id_U \otimes f)) & \\
=\; & id_U \otimes Tr^U(f) & \text{Superposing.}
\end{aligned}
$$

Let us consider the case of $U = V \otimes V$ and $f = c_{V,V}$. Then

$$
\begin{aligned}
c_{V,V} \; =\; & id_{V \otimes V} \otimes Tr^{V \otimes V}(c_{V,V}) & \text{by } (*) \\
=\; & id_{V \otimes V} \otimes Tr^V(id_V) & \text{Vanishing, Yanking} \\
=\; & id_{V \otimes V} & \text{by } (*).
\end{aligned}
$$

$\square$

A.2 Proof of Theorem 7.1.1

From Trace to Fixed Point

Assume that C is a traced cartesian category. From the trace operator Tr, we define an operator $(-)^\dagger$ by

$$
f^\dagger = Tr^X(f; \Delta_X) : A \longrightarrow X
$$

for $f : A \times X \longrightarrow X$. We show that $(-)^{\dagger}$ satisfies conditions 1~4. For ease of calculation, strict associativity is assumed.

Lemma A.2.1 For $f : A \times X \longrightarrow B \times X$,

$$Tr^X(f) = \langle id_A, Tr^X(f;\pi'_{B,X};\Delta_X)\rangle;f;\pi_{B,X} : A \longrightarrow B.$$

<u>Proof of Lemma:</u>

$$
\begin{aligned}
&\quad LHS \\
&= Tr^X(\Delta_{A \times X};(f;\pi_{B,X} \times f;\pi'_{B,X})) \\
&= Tr^X(\Delta_{A \times X};(id_{A \times X} \times f;\pi'_{B,X}));f;\pi_{B,X} && \text{R. Tightening} \\
&= Tr^{A \times X}((id_A \times f;\pi'_{B,X});\Delta_{A \times X});f;\pi_{B,X} && \text{Sliding} \\
&= Tr^{A \times X}((\Delta_A \times f;\pi'_{B,X};\Delta_X);(id_A \times c_{A,X} \times id_X));f;\pi_{B,X} \\
&= \Delta_A;Tr^{A \times X}(id_A \times ((id_A \times f;\pi'_{B,X};\Delta_X);(c_{A,X} \times id_X)));f;\pi_{B,X} && \text{L. Tightening} \\
&= \Delta_A;(id_A \times Tr^{A \times X}((id_A \times f;\pi'_{B,X};\Delta_X);(c_{A,X} \times id_X)));f;\pi_{B,X} && \text{Superposing} \\
&= \langle id_A, Tr^{A \times X}((id_A \times f;\pi'_{B,X};\Delta_X);(c_{A,X} \times id_X))\rangle;f;\pi_{B,X} \\
&= \langle id_A, Tr^A(Tr^X((id_A \times f;\pi'_{B,X};\Delta_X);(c_{A,X} \times id_X)))\rangle;f;\pi_{B,X} && \text{Vanishing} \\
&= \langle id_A, Tr^A(Tr^X(id_A \times f;\pi'_{B,X};\Delta_X);c_{A,X})\rangle;f;\pi_{B,X} && \text{R. Tightening} \\
&= \langle id_A, Tr^A((id_A \times Tr^X(f;\pi'_{B,X};\Delta_X));c_{A,X})\rangle;f;\pi_{B,X} && \text{Superposing} \\
&= \langle id_A, Tr^A(c_{A,A};(Tr^X(f;\pi'_{B,X};\Delta_X) \times id_A))\rangle;f;\pi_{B,X} \\
&= \langle id_A, Tr^A(c_{A,A});Tr^X(f;\pi'_{B,X};\Delta_X)\rangle;f;\pi_{B,X} && \text{R. Tightening} \\
&= \langle id_A, Tr^X(f;\pi'_{B,X};\Delta_X)\rangle;f;\pi_{B,X} && \text{Yanking} \\
&= RHS
\end{aligned}
$$

$\square$

Condition 1:

$$
\begin{aligned}
f^{\dagger} &= Tr^X(f;\Delta_X) \\
&= \langle id_A, Tr^X(f;\Delta_X;\pi'_{X,X};\Delta_X)\rangle;\langle f,f\rangle;\pi_{X,X} && \text{A.2.1} \\
&= \langle id_A, Tr^X(f;\Delta_X)\rangle;f \\
&= \langle id_A, f^{\dagger}\rangle;f.
\end{aligned}
$$

Condition 2:

$$
\begin{aligned}
((g \times id_X);f)^{\dagger} &= Tr^X((g \times id_X);f;\Delta) \\
&= g;Tr^X(f;\Delta) && \text{Left Tightening} \\
&= g;f^{\dagger}.
\end{aligned}
$$

Condition 3:

$$
\begin{aligned}
(f;g)^{\dagger} &= Tr^X(f;g;\Delta) \\
&= Tr^X(f;\Delta;(g \times g)) \\
&= Tr^X(f;\Delta;(id_Y \times g));g && \text{Right Tightening} \\
&= Tr^Y((id_A \times g);f;\Delta);g && \text{Sliding} \\
&= ((id_A \times g);f)^{\dagger};g.
\end{aligned}
$$

Condition 4 (Bekič's lemma):
In the simple slice $C//A$ (page 45), the condition is stated as

$$\langle f,g \rangle^{\dagger} = (\langle id_X,g \rangle;f)^{\dagger};\langle id_X,g \rangle$$

for $f : X \times Y \to X$ and $g : X \times Y \to Y$. By Corollary 6.2.4, $C//A$ is also a traced cartesian category, so it suffices to show this simpler equation from the axioms of traces.

We first calculate

$$
\begin{aligned}
&\quad LHS \\
&= Tr^{X \times Y}(\langle f,g \rangle;\Delta_{X \times Y}) \\
&= Tr^{X}(Tr^{Y}(\langle f,g \rangle;\Delta_{X \times Y})) && \text{Vanish.} \\
&= Tr^{X}(\langle id_X,Tr^{Y}(\langle f,g \rangle;\Delta_{X \times Y};\pi'_{X \times Y \times X,Y};\Delta_Y) \rangle;\langle f,g \rangle;\Delta_{X \times Y};\pi_{X \times Y \times X,Y}) && \text{A.2.1} \\
&= Tr^{X}(\langle id_X,Tr^{Y}(g;\Delta_Y) \rangle;\langle f,g \rangle;\Delta_{X \times Y};\pi_{X \times Y \times X,Y}) \\
&= Tr^{X}(\langle id_X,g^{\dagger} \rangle;\langle f;\Delta_X,g \rangle;(id_X \times c_{X,X})) \\
&= Tr^{X}(\langle \langle id_X,g^{\dagger} \rangle;f;\Delta_X,\langle id_X,g^{\dagger} \rangle;g \rangle;(id_X \times c_{X,Y})) \\
&= Tr^{X}(\langle \langle id_X,g^{\dagger} \rangle;f;\Delta_X,g^{\dagger} \rangle;(id_X \times c_{X,Y})) && \text{Cond.1} \\
&= Tr^{X}(\langle \langle id_X,g^{\dagger} \rangle;f;\Delta_X,id_X \rangle;(id_X \times c_{X,X});(id_X \times g^{\dagger} \times id_X)) \\
&= Tr^{X}(\langle \langle id_X,g^{\dagger} \rangle;f;\Delta_X,id_X \rangle;(id_X \times c_{X,X}));(id_X \times g^{\dagger}) && \text{R.T.}
\end{aligned}
$$

$$
\begin{aligned}
&\quad RHS \\
&= Tr^{X}(\langle id_X,g^{\dagger} \rangle;f;\Delta_X);\langle id_X,g^{\dagger} \rangle \\
&= Tr^{X}(\langle id_X,g^{\dagger} \rangle;f;\Delta_X;(\Delta_X \times id_X));(id_X \times g^{\dagger}) && \text{R.T.} \\
&= Tr^{X}(\langle id_X,g^{\dagger} \rangle;f;\Delta_X;(id_X \times \Delta_X));(id_X \times g^{\dagger})
\end{aligned}
$$

Therefore we have only to show

$$Tr^{X}(\langle \varphi,id_X \rangle;(id_X \times c_{X,X})) = Tr^{X}(\varphi;(id_X \times \Delta_X))$$

where $\varphi = \langle id_{A \times X},g^{\dagger} \rangle;f;\Delta_X : X \longrightarrow X \times X$. This is verified by

$$
\begin{aligned}
&\quad Tr^{X}(\langle \varphi,id_X \rangle;(id_X \times c_{X,X})) \\
&= Tr^{X}(\Delta_X;(\varphi \times id_X);(id_X \times c_{X,X})) \\
&= Tr^{X \times X}((\varphi \times id_X);(id_X \times c_{X,X});(id_{X \times X} \times \Delta_X)) && \text{Sliding} \\
&= Tr^{X}(Tr^{X}((\varphi \times id_X);(id_X \times c_{X,X});(id_{X \times X} \times \Delta_X))) && \text{Vanishing} \\
&= Tr^{X}(\varphi;Tr^{X}((id_X \times c_{X,X});(id_{X \times X} \times \Delta_X))) && \text{L. Tightening} \\
&= Tr^{X}(\varphi;(id_X \times Tr^{X}(c_{X,X};(id_X \times \Delta_X)))) && \text{Superposing} \\
&= Tr^{X}(\varphi;(id_X \times Tr^{X}((\Delta_X \times id_X);(id_X \times c_{X,X});(c_{X,X} \times id_X)))) \\
&= Tr^{X}(\varphi;(id_X \times \Delta_X;Tr^{X}(id_X \times c_{X,X});c_{X,X})) && \text{L.\&R.T.} \\
&= Tr^{X}(\varphi;(id_X \times \Delta_X;(id_X \times Tr^{X}c_{X,X});c_{X,X})) && \text{Superposing} \\
&= Tr^{X}(\varphi;(id_X \times \Delta_X;c_{X,X})) && \text{Yanking} \\
&= Tr^{X}(\varphi;(id_X \times \Delta_X)).
\end{aligned}
$$

From Fixed Point to Trace

Let C be a cartesian category with an operator $(-)^{\dagger}$ which satisfies 1~4. We define

$$Tr^{X}(f) = \langle id_A,(f;\pi'_{B,X})^{\dagger} \rangle;f;\pi_{B,X} : A \longrightarrow B$$

for $f : A \times X \longrightarrow B \times X$, and show that Tr is a trace operator. Again strict associativity is assumed.

Lemma A.2.2 $Tr^X(f) = ((id_A \times \pi'_{B,X}); f)^\dagger; \pi_{B,X}.$

<u>Proof of Lemma:</u>

$$
\begin{aligned}
Tr^X(f) &= \langle id_A, (f; \pi'_{B,X})^\dagger \rangle; f; \pi_{B,X} \\
&= \langle id_A, ((id_A \times \pi'_{B,X}); f)^\dagger \rangle; f; \pi_{B,X} \quad (3) \\
&= ((id_A \times \pi'_{B,X}); f)^\dagger; \pi_{B,X} \quad (1)
\end{aligned}
$$

$\square$

Vanishing:

$$
\begin{aligned}
Tr^1(f) &= \langle id_A, (f; \pi'_{B,1})^\dagger \rangle; f; \pi_{B,1} \\
&= \langle id_A, !_A \rangle; f; id_B \\
&= id_A; f \\
&= f.
\end{aligned}
$$

For $f : A \times X \times Y \longrightarrow B \times X \times Y$, we define

$$
\begin{aligned}
F &= f; \pi_{B \times X, Y}; \pi'_{B,X} &: A \times X \times Y \longrightarrow X \\
G &= f; \pi'_{B \times X, Y} &: A \times X \times Y \longrightarrow Y.
\end{aligned}
$$

Then

$$
\begin{aligned}
(*) \quad Tr^Y(f) &= \langle id_{A \times X}, (f; \pi'_{B \times X, Y})^\dagger \rangle; f; \pi_{B \times X, Y} \\
&= \langle id_{A \times X}, G^\dagger \rangle; f; \pi_{B \times X, Y} &: A \times X \longrightarrow B \times Y.
\end{aligned}
$$

We also note that $f; \pi'_{B, X \times Y} = \langle F, G \rangle : A \times X \times Y \longrightarrow X \times Y$. Then

$$
\begin{aligned}
Tr^{X \times Y}(f) &= \langle id_A, (f; \pi'_{B, X \times Y})^\dagger \rangle; f; \pi_{B, X \times Y} \\
&= \langle id_A, \langle F, G \rangle^\dagger \rangle; f; \pi_{B, X \times Y} \\
&= \langle id_A, \langle id_A, ((id_{A \times X}, G^\dagger); F)^\dagger \rangle; \langle \pi'_{A \times X}, G^\dagger \rangle \rangle; f; \pi_{B, X \times Y} \quad (4) \\
&= \langle id_A, ((id_{A \times X}, G^\dagger); F)^\dagger \rangle; \langle id_{A \times X}, G^\dagger \rangle; f; \pi_{B, X \times Y} \\
&= \langle id_A, (Tr^Y(f); \pi'_{B,X})^\dagger \rangle; Tr^Y(f); \pi_{B,X} \quad (*) \\
&= Tr^X(Tr^Y(f)).
\end{aligned}
$$

Superposing:

$$
\begin{aligned}
& Tr^X((id_A \times c_{C,X}); (f \times g); (id_B \times c_{X,D})) \\
=\ & \langle id_{A \times C}, ((id_A \times c_{C,X}); (f \times g); (id_B \times c_{X,D}); \pi'_{B \times D, X})^\dagger \rangle; \\
& (id_A \times c_{C,X}); (f \times g); (id_B \times c_{X,D}); \pi_{B \times D, X} \\
=\ & \langle id_{A \times C}, ((\pi_{A,C} \times id_X); f; \pi'_{B,X})^\dagger \rangle; (id_A \times c_{C,X}); (f; \pi_{B,X} \times g) \\
=\ & \langle id_{A \times C}, \pi_{A,C}; (f; \pi'_{B,X})^\dagger \rangle; (id_A \times c_{C,X}); (f; \pi_{B,X} \times g) \quad (2) \\
=\ & (\langle id_A, (f; \pi'_{B,X})^\dagger \rangle; f; \pi_{B,X}) \times g \\
=\ & Tr^X(f) \times g.
\end{aligned}
$$

Yanking:

$$
\begin{aligned}
Tr^X(c_{X,X}) &= \langle id_X, (c_{X,X}; \pi'_{X,X})^\dagger \rangle; c_{X,X}; \pi_{X,X} \\
&= \langle id_X, \pi^\dagger_{X,X} \rangle; \pi'_{X,X} \\
&= \pi^\dagger_{X,X} \\
&= \langle id_X, \pi^\dagger_{X,X} \rangle; \pi_{X,X} \qquad (1)\\
&= id_X.
\end{aligned}
$$

Left Tightening:

$$
\begin{aligned}
Tr^X((g \times id_X); f) &= ((id_{A'} \times \pi'_{B,X}); (g \times id_X); f)^\dagger; \pi_{B,X} \qquad A.2.2 \\
&= ((g \times id_{B \times X}); (id_A \times \pi'_{B,X}); f)^\dagger; \pi_{B,X} \\
&= g; ((id_A \times \pi'_{B,X}); f)^\dagger; \pi_{B,X} \qquad (2).
\end{aligned}
$$

Right Tightening:

$$
\begin{aligned}
& Tr^X(f; (g \times id_X)) \\
&= ((id_A \times \pi'_{B,X}); f; (g \times id_X))^\dagger; \pi_{B,X} \qquad A.2.2 \\
&= ((id_A \times g \times id_X); (id_A \times \pi'_{B,X}); f)^\dagger; (g \times id_X); \pi_{B,X} \qquad (3) \\
&= ((id_A \times \pi'_{B',X}); f)^\dagger; \pi_{B',X}; g \\
&= Tr^X(f); g.
\end{aligned}
$$

Sliding:

$$
\begin{aligned}
Tr^X(f; (id_B \times g)) &= ((id_A \times \pi'_{B,X}); f; (id_B \times g))^\dagger; \pi_{B,X} \\
&= ((id_A \times id_B \times g); (id_A \times \pi'_{B,X}); f)^\dagger; (id_B \times g); \pi_{B,X} \qquad (3) \\
&= ((id_A \times \pi'_{B,X'}); (id_A \times g); f)^\dagger; \pi_{B,X'} \\
&= Tr^{X'}((id_A \times g); f).
\end{aligned}
$$

A.3 Proof of Theorem 7.2.1

Let us write $U : S \longrightarrow C$ for the right adjoint of $\mathcal{F}$, and $\varepsilon_X : UX \longrightarrow X$ (in S) for the counit. By definition, we have a natural bijection $(-)^* : S(A, B) \longrightarrow C(A, UB)$. We also define $\theta_{A,X} : A \times UX \longrightarrow U(A \otimes X)$ in C by $\theta_{A,X} = (id_A \otimes \varepsilon_X)^*$. Now we define $(-)^\dagger$ by

$$
f^\dagger = Tr^{UX}(\mathcal{F}(\theta_{A,X}; Uf; \Delta_{UX})); \varepsilon_X : A \longrightarrow X \quad \text{in } S
$$

for $f : A \otimes X \longrightarrow X$ in S.

Condition 1: Following the remark after Theorem 7.2.1, we have only to show the simpler equation $(f; g)^\dagger = (g; f)^\dagger; g$ for $f : X \to Y$ and $g : Y \to X$.

$$
\begin{aligned}
(f; g)^\dagger &= Tr^{UX}(\mathcal{F}(Uf; Ug; \Delta_{UX})); \varepsilon_X \\
&= Tr^{UX}(\mathcal{F}(Uf; \Delta_{UY}; (Ug \times Ug))); \varepsilon_X \\
&= Tr^{UX}(\mathcal{F}(Uf; \Delta_{UY}); (\mathcal{F}Ug \otimes \mathcal{F}Ug))); \varepsilon_X \\
&= Tr^{UX}(\mathcal{F}(Uf; \Delta_{UY}); (id_{UY} \otimes \mathcal{F}Ug)); \mathcal{F}Ug; \varepsilon_X \qquad \text{R. Tightening} \\
&= Tr^{UX}(\mathcal{F}(Uf; \Delta_{UY}); (id_{UY} \otimes \mathcal{F}Ug)); \varepsilon_Y; g \\
&= Tr^{UY}(\mathcal{F}Ug; \mathcal{F}(Uf; \Delta_{UY})); \varepsilon_Y; g \qquad \text{Sliding} \\
&= Tr^{UY}(\mathcal{F}(Ug; Uf; \Delta_{UY})); \varepsilon_Y; g \\
&= (g; f)^\dagger; g.
\end{aligned}
$$

Condition 2:

$$
\begin{aligned}
((\mathcal{F}g \otimes id_X); f)^\dagger &= Tr^{UX}(\mathcal{F}(\theta_{B,X}; U(g \otimes id); Uf; \Delta_{UX})); \varepsilon_X \\
&= Tr^{UX}(\mathcal{F}((g \times id); \theta_{A,X}; Uf; \Delta_{UX})); \varepsilon_X \\
&= Tr^{UX}((\mathcal{F}g \otimes id); \mathcal{F}(\theta_{A,X}; Uf; \Delta_{UX})); \varepsilon_X \\
&= \mathcal{F}g; Tr^{UX}(\mathcal{F}(\theta_{A,X}; Uf; \Delta_{UX})); \varepsilon_X \qquad \text{L. Tightening} \\
&= \mathcal{F}g; f^\dagger.
\end{aligned}
$$

In the proof, we used the naturality of θ:

$$
\theta_{A,X}; U(\mathcal{F}(g) \otimes f) = (g \times Uf); \theta_{B,Y} : A \times UX \longrightarrow U(B \otimes Y)
$$

where $f : X \longrightarrow Y$ in S and $g : A \longrightarrow B$ in C. This is routinely shown as

$$
\begin{aligned}
\theta_{A,X}; U(\mathcal{F}g \otimes f) &= (id_A \otimes \varepsilon_X)^*; (\varepsilon_{A,X}; (\mathcal{F}g \otimes f))^* \\
&= ((id_A \otimes \varepsilon_X)^*; \varepsilon_{A,X}; (\mathcal{F}g \otimes f))^* \\
&= ((id_A \otimes \varepsilon_X); (\mathcal{F}g \otimes f))^* \\
&= (\mathcal{F}g \otimes \varepsilon_X; f)^* \\
&= (\mathcal{F}g \otimes \mathcal{F}((\varepsilon_X; f)^*); \varepsilon_X)^* \\
&= (\mathcal{F}(g \times (\varepsilon_X; f)^*); (id_B \otimes \varepsilon_X))^* \\
&= (g \times (\varepsilon_X; f)^*); (id_B \times \varepsilon_X)^* \\
&= (g \times U(f)); \theta_{B,X}.
\end{aligned}
$$

Remark A.3.1 Alternatively, after establishing the relation with higher-order cyclic sharing theories (or their fragments), we can use the equational theory as an internal language of our structure, and this theorem can be proved by equational reasoning in this language. See Example 7.3.11.   □

A.4   Proof of Proposition 7.1.4

From the First Condition to the Second

Assume that the diagram

$$
\begin{array}{ccc}
A \times X & \xrightarrow{\ f\ } & X \\
{\scriptstyle A \times h}\downarrow & & \downarrow{\scriptstyle h} \\
A \times Y & \xrightarrow[\ g\ ]{} & Y
\end{array}
$$

commutes. Then the following diagram

$$
\begin{array}{ccc}
A \times X & \xrightarrow{f; \Delta; (h \times X)} & Y \times X \\
{\scriptstyle A \times h}\downarrow & & \downarrow{\scriptstyle Y \times h} \\
A \times Y & \xrightarrow[\ g; \Delta\ ]{} & Y \times Y
\end{array}
$$

also commutes. From the uniformity of the trace operator, we have

$$Tr^X(f;\Delta;(h\times X)) = Tr^Y(g;\Delta).$$

By Right Tightening, the left hand side is equal to $Tr^X(f;\Delta);h$. Since $f^\dagger = Tr^X(f;\Delta)$ and $g^\dagger = Tr^Y(g;\Delta)$, we get $f^\dagger;h = g^\dagger$.

From the Second to the First

Assume that the diagram

$$\begin{array}{ccc} A\times X & \xrightarrow{f} & B\times X \\ {\scriptstyle A\times h}\big\downarrow & & \big\downarrow{\scriptstyle B\times h} \\ A\times Y & \xrightarrow[g]{} & B\times Y \end{array}$$

commutes. Then the following diagram

$$\begin{array}{ccc} A\times B\times X & \xrightarrow{(A\times\pi');f} & B\times X \\ {\scriptstyle A\times B\times h}\big\downarrow & & \big\downarrow{\scriptstyle B\times h} \\ A\times B\times Y & \xrightarrow[(A\times\pi');g]{} & B\times Y \end{array}$$

also commutes. Since h satisfies the uniformity condition of the fixed point operator, so does $B\times h$ (by the lemma below). Thus we have

$$((A\times\pi');f)^\dagger;(B\times h) = ((A\times\pi');g)^\dagger.$$

Since $Tr^X(f) = ((A\times\pi');f)^\dagger;\pi$ and $Tr^Y(g) = ((A\times\pi');g)^\dagger;\pi$, we get $Tr^X(f) = Tr^Y(g)$.

Lemma A.4.1 If $h: X \longrightarrow Y$ and $h': X' \longrightarrow Y'$ satisfy the uniformity condition on the fixed point operator, so does $h\times h': X\times X' \longrightarrow Y\times Y'$.

Proof: Assume that the diagram

$$\begin{array}{ccc} A\times X\times X' & \xrightarrow{f} & X\times X' \\ {\scriptstyle A\times h\times h'}\big\downarrow & & \big\downarrow{\scriptstyle h\times h'} \\ A\times Y\times Y' & \xrightarrow[g]{} & Y\times Y' \end{array} \qquad \text{(A.1)}$$

commutes. Our purpose is to show $f^\dagger;(h\times h') = g^\dagger$. By Bekič's lemma, this is equivalent to equations

$$((\langle A\times X, f_2^\dagger\rangle;f_1)^\dagger;h = ((\langle A\times Y, g_2^\dagger\rangle;g_1)^\dagger \qquad \text{(A.2)}$$

$$((\langle A \times X', f'^{\dagger}_1 \rangle; f'_2)^{\dagger}; h' = (\langle A \times Y', g'^{\dagger}_1 \rangle; g'_2)^{\dagger} \tag{A.3}$$

where $f_1 = f; \pi : A \times X \times X' \longrightarrow X$, $f_2 = f; \pi' : A \times X \times X' \longrightarrow X'$, $f'_i = (A \times c_{X',X}); f_i$, and so on. We shall show A.2. A.3 is proved in the same way.

By A.1, the diagrams

$$
\begin{array}{ccc}
A \times X \times X' & \xrightarrow{\ f_1\ } & X \\
{\scriptstyle A \times h \times h'} \downarrow & & \downarrow {\scriptstyle h} \\
A \times Y \times Y' & \xrightarrow[\ g_1\ ]{} & Y
\end{array}
\tag{A.4}
$$

$$
\begin{array}{ccc}
A \times X \times X' & \xrightarrow{\ f_2\ } & X' \\
{\scriptstyle A \times X \times h'} \downarrow & & \downarrow {\scriptstyle h'} \\
A \times X \times Y' & \xrightarrow[{(A \times h \times Y'); g_2}]{} & Y'
\end{array}
\tag{A.5}
$$

commute. From A.5 and the uniformity condition on h',

$$f^{\dagger}_2; h' = ((A \times h \times Y'); g_2)^{\dagger}.$$

By naturality, the right hand side is equal to $(A \times h); g^{\dagger}_2$. Thus we have a commutative diagram

$$
\begin{array}{ccc}
A \times X & \xrightarrow{\ f^{\dagger}_2\ } & X' \\
{\scriptstyle A \times h} \downarrow & & \downarrow {\scriptstyle h'} \\
A \times Y & \xrightarrow[\ g^{\dagger}_2\ ]{} & Y'
\end{array}
\tag{A.6}
$$

From A.4 and A.6,

$$
\begin{array}{ccccc}
A \times X & \xrightarrow{\langle A \times X, f^{\dagger}_2 \rangle} & A \times X \times X' & \xrightarrow{\ f_1\ } & X \\
{\scriptstyle A \times h} \downarrow & & {\scriptstyle A \times h \times h'} \downarrow & & \downarrow {\scriptstyle h} \\
A \times Y & \xrightarrow[\langle A \times Y, g^{\dagger}_2 \rangle]{} & A \times Y \times Y' & \xrightarrow[\ g_1\ ]{} & Y
\end{array}
$$

commutes. Applying the uniformity condition on h, we obtain A.2. $\square$

Remark A.4.2 The corresponding result for the uniformity condition on the trace operator follows trivially from Vanishing. $\square$

A.5 Proof of Proposition 7.2.2

Assume that the diagram

$$
\begin{array}{ccc}
A \otimes X & \xrightarrow{\ f\ } & X \\
{\scriptstyle A \otimes h}\downarrow & & \downarrow{\scriptstyle h} \\
A \otimes Y & \xrightarrow[\ g\ ]{} & Y
\end{array}
$$

commutes. Then the following diagram

$$
\begin{array}{ccccccc}
A \times UX & \xrightarrow{\ \theta_{A,X}\ } & U(A \otimes X) & \xrightarrow{\ Uf\ } & UX & \xrightarrow{\ \Delta_{UX}\ } & UX \times UX \\
{\scriptstyle A \times Uh}\downarrow & & {\scriptstyle U(A \otimes h)}\downarrow & & {\scriptstyle Uh}\downarrow & & \downarrow{\scriptstyle Uh \times Uh} \\
A \times UY & \xrightarrow[\ \theta_{A,Y}\ ]{} & U(A \otimes Y) & \xrightarrow[\ Ug\ ]{} & UY & \xrightarrow[\ \Delta_{UY}\ ]{} & UY \times UY
\end{array}
$$

commutes (in C), which implies that

$$
\begin{array}{ccccc}
A \otimes UX & \xrightarrow{\mathcal{F}(\theta_{A,X};Uf;\Delta_{UX})} & UX \otimes UX & \xrightarrow{\mathcal{F}(Uh)\otimes UX} & UY \otimes UX \\
{\scriptstyle A \otimes \mathcal{F}(Uh)}\downarrow & & & & \downarrow{\scriptstyle UY \otimes \mathcal{F}(Uh)} \\
A \otimes UY & & \xrightarrow[\mathcal{F}(\theta_{A,Y};Ug;\Delta_{UY})]{} & & UY \otimes UY
\end{array}
$$

also commutes. Then, by assumption, we have

$$
Tr^{UX}\left(\mathcal{F}(\theta_{A,X};Uf;\Delta_{UX});(\mathcal{F}(Uh)\otimes UX)\right) = Tr^{UY}\left(\mathcal{F}(\theta_{A,Y};Ug;\Delta_{UY})\right).
$$

The left hand side is equal to $Tr^{UX}\left(\mathcal{F}(\theta_{A,X};Uf;\Delta_{UX})\right);\mathcal{F}(Uh)$ (by Right Tightening). Composing with ε_Y from right to the both sides, we obtain $f^\dagger;h = g^\dagger$. □

Bibliography

[1] Abadi M, Cardelli L, Curien P-L and Levy J-J. Explicit substitutions. Journal of Functional Programming 1991;1(4):375-416

[2] Abramsky S. Retracing some paths in process algebra. In: Proceedings of the 7th International Conference on Concurrency Theory (CONCUR'96). Pisa, Springer Lecture Notes in Computer Science 1119, pp 1-17, 1996

[3] Abramsky S. Axioms for full abstraction and full completeness. Manuscript, LFCS, University of Edinburgh, 1996

[4] Abramsky S, Blute R and Panangaden P. Nuclear and trace ideals in tensored *-categories. To appear in Journal of Pure and Applied Algebra

[5] Aczel P. Non-Well-Founded Sets. CSLI Lecture Notes 14, CSLI Publications, 1988

[6] Aczel P. Replacement systems and the axiomatization of situation theory. In: Proceedings of the First Conference on Situation Theory and Its Applications, Vol.1. Asilomar, CSLI Lecture Notes 22, CSLI Publications, 1990

[7] Ariola ZM and Arvind. Properties of a first-order functional language with sharing. Theoretical Computer Science 1995;146:69-108

[8] Ariola ZM and Blom S. Cyclic lambda calculi. In: Proceedings of the 3rd International Symposium on Theoretical Aspects of Computer Software (TACS'97). Sendai, Springer Lecture Notes in Computer Science 1281, pp 77-106, 1997

[9] Ariola ZM and Felleisen M. A call-by-need lambda calculus. Journal of Functional Programming 1997;7(3):265-301

[10] Ariola ZM and Klop JW. Cyclic lambda graph rewriting. In: Proceedings of the 9th IEEE Symposium on Logic in Computer Science (LICS'94). Paris, IEEE Computer Society Press, pp 416-425, 1994

[11] Ariola ZM and Klop JW. Equational term graph rewriting Fundamentae Infomaticae 1996;26(3,4):207-240

[12] Barber A. Dual intuitionistic linear logic. Technical report ECS-LFCS-96-347, LFCS, University of Edinburgh, 1996

[13] Barber A, Gardner P, Hasegawa M and Plotkin G. From action calculi to linear logic. In: Annual Conference of the European Association for Computer Science Logic (CSL'97), Selected Papers. Arhus, Springer Lecture Notes in Computer Science 1410, pp 78-97, 1998

[14] Barendregt H. The Lambda Calculus: Its Syntax and Semantics. Studies in Logic and the Foundations of Mathematics 103, North Holland, 1984

[15] Barendregt H, van Eekelen M, Glauert J et al. Term graph rewriting. In: Proceedings of the Conference on Parallel Architecture and Langauges Europe (PARLE'87). Springer Lecture Notes in Computer Science 259, pp 141-158, 1987

[16] Barwise J and Moss L. Vicious Circles. CSLI Publications, 1996

[17] Benaissa Z-E-A, Lascanne P and Rose KH. Modeling sharing and recursion for weak reduction strategies using explicit substitution. Rapport de recherche No. 3092, INRIA, 1997

[18] Benton N. A mixed linear non-linear logic: proofs, terms and models. In: 8th Workshop on Computer Science Logic (CSL'94), Selected Papers. Kazimierz, Springer Lecture Notes in Computer Science 933, pp 121-135, 1995

[19] Benton N and Wadler P. Linear logic, monads, and the lambda calculus. In: Proceedings of the 11th IEEE Symposium on Logic in Computer Science (LICS'96). New Brunswick, New Jersey, IEEE Computer Society Press, pp 420-431, 1996

[20] Bierman GM. What is a categorical model of intuitionistic linear logic? In: Proceedings of the 2nd International Conference on Typed Lambda Calculi and Applications (TLCA'95). Edinburgh, Springer Lecture Notes in Computer Science 902, pp 78-93, 1995

[21] Bloom SL and Ésik Z. Iteration Theories. EATCS Monographs on Theoretical Computer Science, Springer-Verlag, 1993

[22] Bloom SL and Ésik Z. Fixed point operators on ccc's. Part I. Theoretical Computer Science 1996;155:1-38

[23] Blute R, Cockett R and Seely RAG. Feedback for linearly distributive categories: traces and fixpoints. To appear in Journal of Pure and Applied Algebra

[24] Braüner T. The Girard translation extended with recursion. In: 8th Workshop on Computer Science Logic (CSL'94), Selected Papers. Kazimierz, Springer Lecture Notes in Computer Science 933, pp 31-45, 1995

[25] Corradini A and Gadducci F. A 2-categorical presentation of term graphs. In: Proceedings of the 7th International Conference on Category Theory and Computer Science (CTCS'97). Santa Margherita Ligure, Genoa, Springer Lecture Notes in Computer Science 1290, pp 87-105, 1997

[26] Crole RL. Categories for Types. Cambridge University Press, 1993

[27] Crole RL and Pitts AM. New foundations for fixpoint computations: FIX-hyperdoctrines and the FIX-logic. Information and Computation 1992;98:171-210

[28] Curien P-L. Categorical Combinators, Sequential Algorithms, and Functional Programming (Second Edition). Birkhäuser, 1993

[29] Day BJ. On closed categories of functors. In: Midwest Category Seminar Reports IV. Springer Lecture Notes in Mathematics 137, pp 1-38, 1970

[30] Eilenberg S and Kelly GM. Closed categories. In: Proceedings of the Conference on Categorical Algebra (La Jolla 1965). Springer-Verlag, pp 421-562, 1966

[31] Filinski A. Recursion from iteration. Lisp and Symbolic Computation 1994;7(1):11-38

[32] Freyd P. Algebraically complete categories. In: Proceedings of 1990 Como Category Theory Conference. Springer Lecture Notes in Mathematics 1144, pp 95-104, 1991

[33] Friedman DP and Felleisen M. The Seasoned Schemer. The MIT Press, 1996

[34] Gardner P. A name-free account of action calculi. In: Proceedings of the 11th International Conference on Mathematical Foundations of Programming Semantics (MFPS'95). New Orleans, Lousiana, Electronic Notes in Computer Science 1, Elsevier, 1995

[35] Gardner P and Hasegawa M. Types and models for higher-order action calculi. In: Proceedings of the 3rd International Symposium on Theoretical Aspects of Computer Software (TACS'97). Sendai, Springer Lecture Notes in Computer Science 1281, pp 583-603, 1997

[36] Girard J-Y. Linear logic. Theoretical Computer Science 1987;50:1-102

[37] Girard J-Y. Geometry of interaction I: interpretation of system F. In: Logic Colloquium '88. North-Holland, pp 221-260, 1989

[38] Hasegawa M. Recursion from cyclic sharing: traced monoidal categories and models of cyclic lambda calculi. In: Proceedings of the 3rd International Conference on Typed Lambda Calculi and Applications (TLCA'97). Nancy, Springer Lecture Notes in Computer Science 1210, pp 196-213, 1997

[39] Hasegawa M. Logical predicates for intuitionistic linear type theories. In: Proceedings of the 4th International Conference on Typed Lambda Calculi and Applications (TLCA'99). L'Aquila, Springer Lecture Notes in Computer Science, 1999

[40] Hasegawa R. The Lagrange-Good inversion as the trace of formal power series. In: Proceedings of the Workshop on Fixed Points in Computer Science (FICS'98). Brno, 1998

[41] Hermida C and Jacobs B. Fibrations with indeterminates: contextual and functional completeness for polymorphic lambda calculi. Mathematical Structures in Computer Science 1995;5:501-531

[42] Hermida C and Power AJ. Fibrational control structures. In: Proceedings of the 6th International Conference on Concurrency Theory (CONCUR'95). Philadelphia, Springer Lecture Notes in Computer Science 962, pp 117-129, 1995

[43] Honda K and Tokoro M. An object calculus for asynchronous communication. In: Proceedings of European Conference on Object-Oriented programming (ECOOP'91). Geneva, Springer Lecture Notes in Computer Science 512, pp 133-147, 1991

[44] Im GB and Kelly GM. A universal property of the convolution monoidal structure. Journal of Pure and Applied Algebra 1986;43:75-88

[45] Jacobs B. Semantics of weakening and contraction. Annals of Pure and Applied Logic 1994;69(1):73-106

[46] Jeffrey A. Premonoidal categories and a graphical view of programs. Manuscript, 1998

[47] Jensen O. A talk at Newton Institute, Cambridge. November 1995

[48] Joyal A and Street R. The geometry of tensor calculus I. Advances in Mathematics 1991;88:55-113

[49] Joyal A and Street R. Braided tensor categories. Advances in Mathematics 1993;102:20-78

[50] Joyal A, Street R and Verity D. Traced monoidal categories. Mathematical Proceedings of the Cambridge Philosophical Society 1996;119(3):447-468

[51] Kassel C. Quantum Groups. Graduate Texts in Mathematics 155, Springer-Verlag, 1995

[52] Kelly GM. Many-variable functorial calculus I. In: Coherence in Categories. Springer Lecture Notes in Mathematics 281, pp 66-105, 1972

[53] Kelly GM. Doctorinal adjunction. In: Proceedings of Sydney Category Theory Seminar, 1972/1973. Springer Lecture Notes in Mathematics 420, pp 257-280, 1974

[54] Kelly GM and Laplaza ML. Coherence for compact closed categories. Journal of Pure and Applied Algebra 1980;19:193-213

[55] Kock A. Strong functors and monoidal monads. Various Publications Series 11, Aarhus Universitet, 1970

[56] Lambek J and Scott PJ. Introduction to Higher Order Categorical Logic. Cambridge University Press, 1986

[57] Launchbury J. A natural semantics for lazy evaluation. In: Proceedings of the 21st ACM Symposium on Principles of Programming Languages (POPL'93). Charleston, ACM Press, pp 144-154, 1993

[58] Lawvere FW. Functorial semantics of algebraic theories. Proceedings of National Academy of Science 1963;50:869-872

[59] Mac Lane S. Categories for the Working Mathematician. Graduate Texts in Mathematics 5, Springer-Verlag, 1971

[60] Maraist J, Odersky M, Turner D and Wadler P. Call-by-name, call-by-value, call-by-need, and the linear lambda calculus. In: Proceedings of the 11th International Conference on Mathematical Foundations of Programming Semantics (MFPS'95). New Orleans, Lousiana, Electronic Notes in Computer Science 1, Elsevier, 1995

[61] Mifsud A. Control Structures. PhD thesis, LFCS, University of Edinburgh, 1996

[62] Mifsud A, Milner R and Power AJ. Control structures. In: Proceedings of the 10th IEEE Symposium on Logic in Computer Science (LICS'95). San Diego, IEEE Computer Society Press, pp 188-198, 1995

[63] Milner R. Communications and Concurrency. Prentice Hall, 1989

[64] Milner R. The polyadic π-calculus: a tutorial. In: Logic and Algebra of Specification. Springer-Verlag, pp 203-246, 1992

[65] Milner R. Functions as processes. Mathematical Structures in Computer Science 1992;2:119-141

[66] Milner R. Higher-order action calculi. In: 7th Workshop on Computer Science Logic (CSL'93), Selected Papers. Swansea, Springer Lecture Notes in Computer Science 832, pp 238-260, 1994

[67] Milner R. Action calculi V: reflexive molecular forms (with Appendix by O. Jensen). Manuscript, LFCS, University of Edinburgh, 1994

[68] Milner R. Calculi for interaction. Acta Informatica 1996;33(8):707-737

[69] Milner R, Parrow J and Walker D. A calculus of mobile processes, part I + II. Information and Computation 1992;100(1):1-77

[70] Miyoshi H. Rewriting logic for cyclic sharing structures. In: Proceedings of the 3rd Fuji International Symposium on Functional and Logic Programming (FLOPS'98). Kyoto, World-Scientific, pp 167-186, 1998

[71] Moggi E. Computational lambda-calculus and monads. Technical report ECS-LFCS-88-66, LFCS, University of Edinburgh, 1988

[72] Moggi E. Notions of computation and monads. Information and Computation 1991;93:55-92

[73] Moggi E. Metalanguages and applications. In: Semantics and Logic of Computation. Cambridge University Press, 1997

[74] Plotkin G and Simpson A. Properties of Fixed Points in Axiomatic Domain Theory. In: Proceedings of the Workshop on Fixed Points in Computer Science (FICS'98). Brno, 1998

[75] Power AJ. Elementary control structures. In: Proceedings of the 7th International Conference on Concurrency Theory (CONCUR'96). Pisa, Springer Lecture Notes in Computer Science 1119, pp 115-130, 1996

[76] Power AJ. Premonoidal categories as categories with algebraic structure. Manuscript, 1996

[77] Power AJ and Robinson EP. Premonoidal categories and notions of computation. Mathematical Structures in Computer Science1997;7(5):453-468

[78] Reshetikhin NYu and Turaev VG. Ribbon graphs and their invariants derived from quantum groups. Communications in Mathematical Physics 1990;127:1-26

[79] Selinger P. Order-incompleteness and finite lambda models. In: Proceedings of the 11th IEEE Symposium on Logic in Computer Science (LICS'96). New Brunswick, New Jersey, IEEE Computer Society Press, pp 432-439, 1996

[80] Shum MC. Tortile tensor categories. Journal of Pure and Applied Algebra 1994;93:57-110

[81] Simpson AK. Recursive types in Kleisli categories. Manuscript, LFCS, University of Edinburgh, 1992

[82] Simpson AK. A characterisation of the least-fixed-point operator by dinaturality. Theoretical Computer Science 1993;118:301-314

[83] Simpson AK. Categorical completeness results for the simply-typed lambda-calculus. In: Proceedings of the 2nd International Conference on Typed Lambda Calculi and Applications (TLCA'95). Edinburgh, Springer Lecture Notes in Computer Science 902, pp414-427, 1995

[84] Sleep M, Plasmeijer M and van Eekelen M (eds.). Term Graph Rewriting: Theory and Practice. John Wiley & Sons, 1993

[85] Street R and Dubuc E. Dinatural transformations. In: Springer Lecture Notes in Mathematics 137, pp 126-137, 1970

[86] Thielecke H. Continuation passing style and self-adjointness. In: Proceedings of the 2nd ACM SIGPLAN Workshop on Continuations (CW'97). Paris, BRICS Notes Series NS-96-13, 1996

[87] Turner DA. A new implementation technique for applicative languages. Software – Practice and Experience 1979;9:31-49

Index